Student Study Guide

Jay Templin
La Salle University

Introduction to
BIOLOGY

SYLVIA S. MADER

WCB Wm. C. Brown Publishers
Dubuque, Iowa • Melbourne, Australia • Oxford, England

Wm. C. Brown Communications, Inc.

President and Chief Executive Officer *G. Franklin Lewis*
Corporate Senior Vice President, President of WCB Manufacturing *Roger Meyer*
Corporate Senior Vice President and Chief Financial Officer *Robert Chesterman*

Copyright © 1994 by Wm. C. Brown Communications, Inc. All rights reserved

A Times Mirror Company

ISBN 0-697-16879-4

No part of this publication may be reproduced, stored in a retrieval system, or transmitted, in any form or by any means, electronic, mechanical, photocopying, recording, or otherwise, without the prior written permission of the publisher.

Printed in the United States of America by Wm. C. Brown Communications, Inc., 2460 Kerper Boulevard, Dubuque, IA 52001

10 9 8 7 6 5 4 3 2 1

CONTENTS

PART I THE CELL

1 Characteristics and Study of Life
2 The Chemical Basis of Life
3 Structure and Function of Cells
4 Cellular Energy

PART II GENETIC BASIS OF LIFE

5 Cell Reproduction
6 Mendelian Genetics
7 DNA and Biotechnology

PART III EVOLUTION AND DIVERSITY

8 Charles Darwin and the Process of Evolution
9 Origin of Life
10 Viruses and Kingdoms Monera, Protista, and Fungi
11 Plant Kingdom
12 Animal Kingdom

PART IV PLANT STRUCTURE AND FUNCTION

13 Plant Organization and Growth
14 Plant Physiology and Reproduction

PART V ANIMAL STRUCTURE AND FUNCTION

15 Animal Organization and Homeostasis
16 Circulation, Blood, and Immunity
17 Digestion, Respiration, and Excretion
18 Nervous and Endocrine Systems
19 Animal Reproduction and Development

PART VI BEHAVIOR AND ECOLOGY

20 Animal Behavior
21 The Biosphere and Ecosystems
22 Population and Environmental Concerns

INTRODUCTION TO THE STUDENT

This study guide contains a number of features to help you learn the key facts and concepts of biology found in each of the textbook chapters. These features are listed here.

Chapter Concepts: What should you be able to do when you have mastered the content of the chapter? The list of objectives tells you. Read these study objectives before reading the chapter. This will prepare you for what you can accomplish while studying the textbook chapter and using the corresponding study guide sections.

Chapter Review: Read this summary of the textbook chapter before studying that chapter. By doing this you will gain a quick overview of the content of the chapter before studying it in depth.

Vocabulary: An extensive vocabulary is available to describe the phenomena from the world of biology. This section offers you practice to learn some of the major terms emphasized in the chapter. Note that a page reference from the textbook is provided next to each listed term in this section of the study guide, allowing you to locate the term and learn about its meaning. Using this information you can match the listed terms to their meanings that are also listed in this section. While looking up these terms and studying them, you are also starting to study the chapter in more detail.

Learning Activities: These activities, with textbook page references, flow directly from the study objectives. The activities make up most of each study guide chapter. Therefore, they will require most of your time and attention. Each set of activities challeges you to read the appropriate section in the textbook chapter, applying information from the text as you answer questions from the activities. Through this process you will explore the chapter in depth and master its content as indicated by the study objectives.

Critical Thinking: The several questions in this section ask you to apply the facts and concepts of biology, taking it a step further than it was presented in the chapter. You must think creatively and often make a deduction.

Chapter Test: Have you mastered the chapter content? The questions in this concluding section allow you to make that assessment. Answer the questions and then check the answer key to compute the percentage that you have answered correctly.

CHAPTER 1

Characteristics and the Study of Life

CHAPTER CONCEPTS

1. Life has levels of organization less complex than the individual and more complex than the individual.
2. Organisms must take in nutrients and energy from the environment to maintain their organization.
3. During reproduction an organizational blueprint is passed on to offspring; after development the blueprint is made known.
4. Members of a species have adaptations to a particular environment.
5. Evolution accounts for the unity (sameness) and diversity of life.
6. The scientific method is appropriate for gathering data and making conclusions about the natural world.

CHAPTER REVIEW

Although life holds tremendous diversity, there are also many common characteristics among living forms. For example, living things show a high degree of organization. This organization is exhibited at the smallest unit of life, the cell. Although some organisms consist of only one cell, other living things are multicellular. In these organisms, similar cells compose tissues, tissues form organs, and organs are part of organ systems. Living things also need a constant input of energy, which is used for various metabolic processes. Within an ecosystem, there is a flow of energy and a cycling of nutrients. Organisms maintain a constant internal environment through homeostasis. They also respond to external stimuli and reproduce. All of these common characteristics represent unity among living organisms. The process of evolution can explain both the unity and the diversity of life.

Most biologists recognize five kingdoms of living things. The kingdom is the broadest, most general taxonomic level. The species is the most exact level of classification.

To learn more about the natural world, scientists use the scientific method. This provides a series of steps and controlled experiments, generating systematic answers to scientifically based questions.

VOCABULARY

The list in the first column includes some of the chapter's terms. Each term is followed by the page where it first appears. Locate each term in the chapter and read its description. Then match the meanings to the terms.

1. adaptation (p. 6) G
2. cell (p. 4) B
3. genus (p. 5) D
4. homeostasis (p. 6) E

a. approach to gather information
b. basic unit of life
c. chemical energy transformation
d. group of species

5. hypothesis (p. 8) H
6. metabolism (p. 6) C
7. scientific method (p. 8) A
8. species (p. 4) F

e. maintenance of constancy
f. members commonly reproduce
g. modification promoting survival
h. step formulated from data

Answers: 1. g 2. b 3. d 4. e 5. h 6. c 7. a 8. f

LEARNING ACTIVITIES

Study the text section by section as you answer the following questions.

Characteristics of Life (p. 6)

1. Match the characteristics of life with the descriptions that follow.

    1. Living things are organized.
    2. Living things respond.
    3. Living things metabolize.
    4. Living things reproduce.
    5. Living things can adapt.

    a. Frogs have a life cycle that includes an egg, a larva (tadpole) that undergoes the process of metamorphosis, and an adult. REPRODUCE
    b. Humans immediately remove their hands from a hot object. RESPOND
    c. All living things are composed of cells. ORGANIZED
    d. A flounder is a flattened fish that lives on the bottom of bodies of water, while a tuna is a streamlined fish that swims in the open sea. ADAPT
    e. Most cells break down the sugar glucose to carbon dioxide and water. METABOLIZE

2. a. The following states levels of biological organization randomly. List them in order, from smallest to largest.
tissue, organ, cell, organ system, population, organism, ecosystem   PG 4   FIG 1.1

1. CELL (smallest)   2. TISSUE   3. ORGAN
4. ORGAN SYSTEM   5. ORGANISM   7. ECOSYSTEM (largest)
6. POPULATION

b. Among the levels of biological organization, at which level do the properties of life first emerge?
c. At which level do different groups of organisms interact with their environment?

Classification of Organisms (p. 4)

3. The following states the levels of classification randomly. List them in order, from smallest (most exact) to largest (most general).
phylum, family, order, kingdom, genus, class, species   P6.5

KINGDOM (smallest)   CLASS   GENUS
PHYLUM   ORDER   SPECIES (largest)
   FAMILY

4. Name the kingdom described in each of the following statements.   PG 5   FIG. 1.2

ANIMALIA a. ingest food
FUNGI b. absorb food—molds and mushrooms
PLANTAE c. photosynthesize food—include ferns
PROTISTA d. include protozoa and algae
MONERA e. absorb food—include bacteria

Unity and Diversity of Life (p. 7)

5. Label each of the following statements as an example of the unity or the diversity of life.

   DIVERSITY a. Fungi absorb food; plants carry out photosynthesis.
   UNITY b. Homeostasis, metabolism, and evolution are characteristics of living things.
   UNITY c. Life began with single cells.
   UNITY d. Living things consist of cells.
   DIVERSITY e. Maple trees have broad, flat leaves and pine trees have leaves that resemble needles.

The Process of Science (p. 13)

6. Complete the blanks in the following paragraph with the correct steps of the scientific method.

Initially, data are used to formulate a(n) (a) HYPOTHESIS . This becomes the basis for new (b) OBSERVATIONS . A(n) (c) THEORY can be constructed when many other experiments arrive at similar conclusions. Scientific experiments usually have a(n) (d) CONTROL group. PG. 8

Answers:
1. a. 4 b. 2 c. 1 d. 5 e. 3
2. a. cell (smallest), tissue, organ, organ system, organism, population, ecosystem (largest) b. cell c. ecosystem
3. species (smallest), genus, family, order, class, phylum, kingdom (largest)
4. a. Animalia b. Fungi c. Plantae d. Protista e. Monera
5. a. diversity b. unity c. unity d. unity e. diversity
6. a. hypothesis b. observation c. theory d. control

CRITICAL THINKING QUESTIONS

1. Why is any level of organization not a mere sum of its parts?
2. What indications show diversity among the uniform properties of living things?

Answers:
1. At each level, there are emergent properties not seen at the lower levels of organization. The cell, for example, is the basic living unit, which is more than just a combination of chemicals.
2. Each kind of organism, a species, has unique adaptations resulting from its evolution. Members of the kingdom of animals, for example, differ in body forms and behavior patterns.

CHAPTER TEST

Indicate your answers by circling the letter. Do not refer to the text when taking this test.

1. The binomial name *Pisum sativum* refers to the taxonomic levels of
   a. class and order. ↑Genus ↑species
   (b) genus and species.
   c. kingdom and phylum.
   d. order and phylum.
2. Select the taxonomic level that is a group of species.
   a. class
   (b) genus
   c. order
   d. phylum
3. Select the largest, broadest taxonomic level among these choices.
   a. class
   b. order
   (c) kingdom
   d. phylum

4. Which is *not* a general characteristic of life?
   a. ability to respond
   b. metabolism
   c. organization
   ⓓ positive feedback
5. The smallest level of organization where the characteristics of life emerge is the
   a. atom.
   ⓑ cell.
   c. molecule.
   d. population.
6. The term metabolism refers to
   ⓐ chemical energy transformation.
   b. maintenance of internal conditions.
   c. the ability to respond to stimuli.
   d. the lack of reproduction.
7. Taxonomists
   ⓐ classify organisms.
   b. study cells under the microscope.
   c. dissect organisms.
   d. observe the planets.
8. An adaptation
   a. produces extinction.
   ⓑ promotes survival.
   c. destroys ecosystems.
   d. represents a habitat.
9. Select the step of the scientific method produced from the accumulation of data.
   ⓐ hypothesis
   b. law
   c. observation
   d. theory
10. Scientific methods usually have a(n) _____ group.
    a. abiotic
    b. biotic
    ⓒ control
    d. observational

Answers: 1. b 2. b 3. c 4. d 5. b 6. a 7. a 8. b 9. a 10. c

CHAPTER

2

The Chemical Basis of Life

CHAPTER CONCEPTS

1. Living and nonliving things are matter and obey the same physical and chemical laws.
2. The atoms and bonds within a molecule determine its physical and chemical properties.
3. Life is dependent on properties of water, an inorganic compound.
4. The properties of organic macromolecules (carbohydrates, lipids, proteins, and nucleic acids) determine the roles they play in the cells and in the body of an organism.

CHAPTER REVIEW

Life has a chemical basis. All matter, living or nonliving, consists of simple substances called elements. An element is composed of discrete units called atoms. The atom of each kind of element is unique by the definite number and arrangements of 3 subatomic particles: protons, neutrons, and electrons. The electrons are located in energy shells around the nucleus. Atoms fulfill a stable outer shell electron configuration by either sharing or transferring electrons with other atoms. Through these processes, they form bonds. Two major kinds of bonds are formed: ionic and covalent. Ionic bonds result from the transfer of electrons; covalent bonds result from electron sharing.

Water is an important biological compound, making life possible on earth. As a solvent, water tends to promote the ionization of substances dissolved in it. Acids and bases are types of compounds that dissolve in water, determining the pH of a solution. The acidity of a system is indicated through the pH scale.

The unique properties of carbon permit the formation of many kinds of organic molecules. At the molecular level, this variety accounts for the diversity of living things.

Several classes of organic molecules have biological importance. One of these, the carbohydrates, consist of several subclasses: monosaccharides, disaccharides, and polysaccharides. The monosaccharides and disaccharides, which are sugars, provide an immediate energy source for organisms. Some polysaccharides store energy, whereas others contribute structurally.

Fatty acids and glycerol are the building blocks of triglycerides. Fatty acids may be either saturated or unsaturated. Triglycerides store energy efficiently. Phospholipids differ in some of their components compared to triglycerides. These structural differences endow these molecules with different biological abilities. Phospholipids, for example, are a major component of plasma membrane structure and determine its properties. Steroids are another type of lipid with several biological roles, including serving as hormones.

Proteins have a variety of biological roles, ranging from transport to structural functions. The subunits of these macromolecules are amino acids. The amino acids are joined by peptide bonds in protein molecules.

DNA, RNA, and ATP are nucleic acids. The structural units of these molecules are nucleotides. DNA and RNA differ structurally in several ways. DNA makes up the genes in cells. RNA directs the synthesis of proteins in a cell. ATP is a universal energy currency used in cells.

VOCABULARY

The list in the first column includes some of the chapter's terms. Each term is followed by the page where it first appears. Locate each term in the chapter and read its description. Then match the meanings to the terms.

_d_	1. acid (p. 17)	a.	simplest substance of matter
_G_	2. base (p. 17)	b.	electrons are shared
_B_	3. covalent bond (p. 16)	c.	electrons are transferred
_E_	4. electron (p. 14)	d.	lowers the pH
_A_	5. element (p. 14)	e.	negative particle in atom
_C_	6. ionic bond (p. 15)	f.	positive particle in atom
_H_	7. protein (p. 20)	g.	raises the pH
_F_	8. proton (p. 14)	h.	chains of amino acids

Answers: 1. d 2. g 3. b 4. e 5. a 6. c 7. h 8. f

LEARNING ACTIVITIES

Study the text section by section as you answer the following questions.

Atoms (p. 14)

1. Name the 3 most stable subatomic particles in an atom: __P__, __N__, and __e__.
2. An element has 9 protons and 9 neutrons in each of its atoms.
   a. Its atomic number = __9__.
   b. Its atomic weight = __18__.
   c. The number of electrons in this atom = __9__.

Molecules (p. 15)

3. Label each of the following statements as describing covalent bonding (C) or ionic bonding (I).
   __I__ a. Electrons are transferred between atoms.
   __C__ b. Electrons are shared between atoms.
   __C__ c. This bond is formed in sodium chloride.
   __I__ d. This bond is formed in the oxygen molecule.
   __C__ e. This bond forms within a water molecule.
   __I__ f. Compounds with this type of bonding dissociate in solution.

Water (p. 16)

4. In each of the following pairs of statements, indicate which one correctly describes how hydrogen bonding affects the properties of water. Hydrogen bonding causes water

   __2__ a. 1. to boil at a lower temperature than expected.
             2. to boil at a higher temperature than expected.
   __2__ b. 1. to be more dense as ice than as liquid water.
             2. to be less dense as ice than as liquid water.
   __1__ c. 1. to absorb heat with a minimal change in temperature.
             2. to absorb heat with a maximum change in temperature.
   __1__ d. 1. to be cohesive—the water molecules cling to one another.
             2. to lack cohesiveness—the water molecules repel one another.

5. Refer to the chemical properties of water when answering the following questions.
   a. What makes water a good solvent?

   b. How does water moderate temperature?

c. What allows ice to float on liquid water?
IT'S DENSITY

Acids and Bases (p. 18)

6. Label each of the following statements as describing an acid (A) or a base (B).
   - _A B_ a. They take up hydrogen ions in solution.
   - _A_ b. HCl is an example.
   - _B_ c. NaOH is an example.
   - _A_ d. They release hydrogen ions in solution.
   - _A_ e. They lower the pH.
   - _B_ f. They raise the pH.
   - _A_ g. They produce a pH of 6 in solution.
   - _B_ h. They produce a pH of 8 in solution.

Carbohydrates (p. 18)

7. Complete the following table:

Carbohydrate	Subunit(s)	Biological Function(s)
sucrose	GLUCOSE, FRUCTOSE	TRANSPORT SUGAR IN PLANTS
lactose	GALACTOSE, LACTOSE	ENERGY SOURCE IN MILK
starch	GLUCOSE	'' STORAGE IN PLANTS
glycogen	GLUCOSE	'' '' '' ANIMALS
cellulose	GLUCOSE	PLANT STRUCTURE

8. From the table in #7, list those that are
   a. monosaccharides. GLUCOSE, FRUCTOSE, GALACTOSE
   b. disaccharides. MALTOSE, LACTOSE, SUCROSE
   c. polysaccharides. STARCH, GLYCOGEN, CELLULOSE

9. The following questions relate to this diagram.
   a. Write the molecular formula for each molecule shown. $C_6H_{12}O_6$
   b. The molecule on the left is the __OPEN__ form of the molecule.
   c. The molecule on the right is the __RING__ form of the molecule.

$C_6H_{12}O_6$ = GLUCOSE

a. Open-chain form of glucose
b. Ring form of glucose

Lipids (p. 19)

10. Complete the following table:

Macromolecule	Biological Function
triglycerides	LONG TERM ENERGY STORAGE COMPOUND.
phospholipids	WHEN PLACED IN H₂O FORM A SHEET IN WHICH THE POLAR HEAD FACE OUTWARDS, NONPOLAR FACE EACH OTHER.
steroids	SERVE AS HORMONES OR CHEMICAL MESSENGERS

11. Write the word saturated or unsaturated beneath each structure.

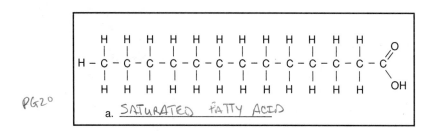

PG20 a. SATURATED FATTY ACID

b. UNSATURATED FATTY ACID

Proteins (p. 20) P 21

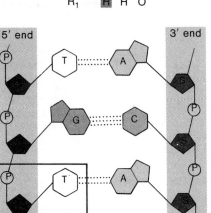

12. a. What type of bond is shown here? PEPTIDE BOND
    b. This bond unites 2 __AMINO ACIDS__.
    c. These 2 bonded structures are the subunits of a(n) __POLYPEPTIDE BOND__.
    d. A sequence of these bonded structures determines the __PRIMARY__ structure of the molecule.
    e. What are the other levels of molecule structure?

Nucleic Acids (p. 22)

13. Study this representation of a nucleic acid.

    PG 22 a. Which molecules make up the backbone of the nucleic acid in the figure? BONDED SUGAR - PHOSPHATE molecules
    b. Which molecules project to one side of the upright portion of this molecule? NITROGEN BASE

    RNA THYMINE c. If a nucleic acid is RNA, state three ways SINGLE CHAIN
    IS REPLACED BY that it differs from DNA. DOUBLE CHAIN, RNA SUGAR IS RIBOSE
    URACIL.
    d. What are the functions of DNA and RNA?
       DNA IS PASSED FROM 1 GEN TO THE NEXT.

(T) THYMINE IS PAIRED W/ (A) ADENINE AND (G) GUANINE
IS PAIRED W/ (C) CYTOSINE. THIS IS KNOWN AS
COMPLEMENTARY BASE
PAIRING.

HYDROGEN BONDS
HOLD THE CHAIN
TOGETHER.

URACIL (YUR-UH-SILL)

LADDER STRUCTURE
DNA

Answers:
1. proton, neutron, electron
2. a. 9 b. 18 c. 9
3. a. I b. C c. I d. C e. C f. I
4. a. 2 b. 2 c. 1 d. 1
5. a. The positive and negative ends of the water molecule attract and disperse charged particles.
   b. It takes up and releases large amounts of heat energy without a change in temperature.
   c. Water is most dense at 4° Centigrade, and it expands if the temperature drops from that point. It is less dense at 0° Centigrade, the temperature of ice.
6. a. B b. A c. B d. A e. A f. B g. A h. B
7. | **Subunit(s)** | **Biological Function(s)** |
   |---|---|
   | glucose, fructose | transport sugar in plants |
   | glucose, galactose | energy source in milk |
   | glucose | energy storage in plants |
   | glucose | energy storage in animals |
   | glucose | plant structure |
8. a. glucose, fructose, galactose
   b. maltose, lactose, sucrose
   c. starch, glycogen, cellulose
9. a. $C_6H_{12}O_6$ for each b. open-chain c. ring
10. triglyceride—long-term energy storage
    phospholipid—protective covering to water loss
    steroid—hormone and plasma membrane structure
11. saturated fatty acid is on the (left); unsaturated fatty acid is on the (right)
12. a. peptide b. amino acids c. protein or polypeptide d. primary e. secondary, tertiary, and quaternary
13. a. bonded sugar-phosphate molecules
    b. nitrogen bases
    c. The sugar is ribose. One of the nitrogen bases is uracil instead of thymine. RNA is usually a single chain of nucleotides.
    d. They function in the storage and transfer of information in genetic systems.

CRITICAL THINKING QUESTIONS

1. Element X has an atomic number of 4, whereas element Y has an atomic number of 18. Which element is much more reactive and why?
2. What are the similarities and differences between glycogen and starch?

Answers:
1. The electron arrangement in X is 2 – 2. It has only 2 electrons in its outer shell. The electron arrangement in Y is 2 – 8 – 8. It has 8 electrons in its outer shell.
2. Both glycogen and starch are polysaccharides with glucose as the subunit. Both store energy, but starch is found in plants and glycogen is found in some animals. Glycogen is a branching molecule, while starch exhibits very little branching.

CHAPTER TEST

Indicate your answers by circling each letter. Do not refer to the text when taking this test.

1. An element has an atomic number of 11 and an atomic weight of 23. Its number of neutrons is
   a. 11.
   b. 12.
   c. 23.
   d. 24.
2. The atom of an element has 1 proton and 2 neutrons. Its atomic number is
   a. 1.
   b. 2.
   c. 3.
   d. 4.
3. The atom of an element has 6 protons and 8 neutrons. Its number of electrons is
   a. 6.
   b. 8.
   c. 12.
   d. 14.
4. An atom has 11 electrons and 12 neutrons. Its atomic weight is
   a. 1.
   b. 11.
   c. 12.
   d. 23.

5. Which is *not* a property of water?
   a. easily changed from liquid to gas ✓
   b. good solvent
   c. maximum density at 4° C
   d. molecules are cohesive
6. Select the *incorrect* statement about acids.
   a. They dissolve in a solution.
   b. They donate hydrogen ions in solution.
   c. HCl is an example.
   d. They tend to raise the pH. ✓
7. Select the *incorrect* statement about bases.
   a. They can be dissolved in solution.
   b. They furnish hydroxide ions in solution.
   c. NaOH is an example.
   d. They tend to lower the pH. ✓

8–15. Match the 4 main classes of organic molecules (right) with the descriptive phrases (left).

Answer	#	Description		Class
A	8.	sucrose is a member	a.	carbohydrates
B	9.	glycerol is a component	b.	fats
d	10.	direct the synthesis of proteins	c.	proteins
d	11.	contain the base uracil	d.	nucleic acids
c	12.	collagen is a member		
b	13.	triglycerides are members		
b	14.	unsaturated subunits		
c	15.	some have enzymatic roles		

Answers: 1. b 2. a 3. a 4. d 5. a 6. d 7. d 8. a 9. b 10. d 11. d 12. c 13. b 14. b 15. c

CHAPTER 3

Structure and Function of Cells

CHAPTER CONCEPTS

1. Life begins at the cellular level and cells carry out life's functions.
2. All organisms are made up of cells which arise only from preexisting cells.
3. Cells are compartmentalized and highly organized.
4. Cells vary in structure and have specific functions in organisms.
5. A form-function relationship pertains to organelles, cells, and other levels of organization.

CHAPTER REVIEW

Cells are the smallest units displaying the properties of life. Cells normally have a sufficient amount of plasma membrane to serve the cytoplasm. Several studies led to the establishment of the fluid-mosaic model of plasma membrane structure. All cells are bound by this plasma membrane, which allows only certain molecules to freely enter and exit the cytoplasm. In plants, a cell wall is external to the plasma membrane.

All organisms are composed of cells. There are two major kinds of cells—prokaryotic and eukaryotic. They differ by the organization of chromosomal DNA and the number of organelles in the cytoplasm.

Molecules move across cell membranes several ways: diffusion, osmosis, facilitated transport, active transport, endocytosis, and exocytosis.

Cells of plants and animals have a particular form and function. Their cells, which are eukaryotic, contain a variety of organelles in the cytoplasm. Ribosomes are the site of protein synthesis. They may exist freely or be attached to the endoplasmic reticulum. Several structures are part of the endomembrane system of the cell. The endoplasmic reticulum provides channels for transport of substances through the cell. Substances are packaged and stored in the Golgi apparatus. Lysosomes contain enzymes promoting the breakdown of cell substances.

Some organelles are specialized to handle energy. Chloroplasts are the site of photosynthesis, whereas the mitochondrion is the region for aerobic cellular respiration.

The cytoskeleton contains actin filaments, intermediate filaments, and microtubules. They function to maintain cell shape and the movement of cell parts.

VOCABULARY

The list in the first column incudes some of the chapter's terms. Each term is followed by the page where it first appears. Locate each term in the chapter and read its description. Then match the meanings to the terms.

J	1. chloroplast (p. 33)	a.	control center
F	2. diffusion (p. 30)	b.	movement of water through a membrane
I	3. ER (p. 33)	c.	site of protein synthesis
H	4. lysosome (p. 33)	d.	general membrane-enclosed space
E	5. microtubule (p. 37)	e.	part of cytoskeleton
A	6. nucleus (p. 32)	f.	general spreading out of molecules
B	7. osmosis (p. 31)	g.	separates cell interior from the outside
G	8. plasma membrane (p. 29)	h.	contains hydrolytic enzymes
C	9. ribosome (p. 33)	i.	smooth or rough
D	10. vacuole (p. 36)	j.	site of photosynthesis

Answers: 1. j 2. f 3. i 4. h 5. e 6. a 7. b 8. g 9. c 10. d

LEARNING ACTIVITIES

Study the text section by section as you answer the following questions.

Plasma Membrane (p. 29)

1. Three of the following 5 statements reflect conclusions of studies leading to the creation of the fluid-mosaic model. Check the 3 correct statements.
   - ✓ a. Channels in the sandwich model of membrane structure allow passage of polar substances.
   - ✓ b. Proteins are part of the membrane.
   - c. The plasma membrane lacks phospholipids.
   - d. Membranes in the cell have varying composition.
   - ✓ e. Proteins make up the outside and inside layers by the sandwich model of membrane structure.

Permeability (p. 30)

2. Label each of the following as describing active transport, diffusion, or osmosis.
   - a. O _____ Algae in a pond become dehydrated.
   - b. O _____ A hypertonic solution draws water.
   - c. O _____ A red blood cell bursts in a person's bloodstream.
   - d. AT _____ ATP is required by cells.
   - e. D _____ Dye crystals spread out in a beaker of water.
   - f. AT _____ Marine fish use energy to expel salt through their gills.
   - g. D _____ Perfume is sensed from the other side of a room.
   - h. AT _____ Plant root cells work to extract inorganic ions from the soil.

3. Answer the following questions about solutions with the correct percentages or the terms hypertonic, hypotonic, or isotonic.
   - a. If a solution is 8% solute, it is _____92_____ % solvent.
   - b. If a solution is 99.5% solvent, it is _____0.5_____ % solute.
   - c. If solution A is 2% solute and solution B is 3% solute, then solution A is __HYPO__ and solution B is __HYPER__.
   - d. Compared to solution A in statement c, a solution with 2% solute is __ISO__.

4. Label each of the following as describing facilitated transport only, active transport only, or both processes.
   - a. __BOTH__ Uses a carrier molecule.
   - b. __FD__ Substances travel from higher to lower concentration.
   - c. __AT__ Substances travel from lower to higher concentration.
   - d. __AT__ Energy is required.
   - e. __FD__ Energy is not required.

Eukaryotic Cell Organelles (p. 32)

5. Select the correct phrases pertaining to eukaryotic cells.
   a. found in plants and animals ✓
   b. contain organelles ✓
   c. nucleus is not well defined
   d. organelles are present ✓
   e. found in bacteria
6. For a–j in the figure of the animal cell, name the organelles and state the function of each.

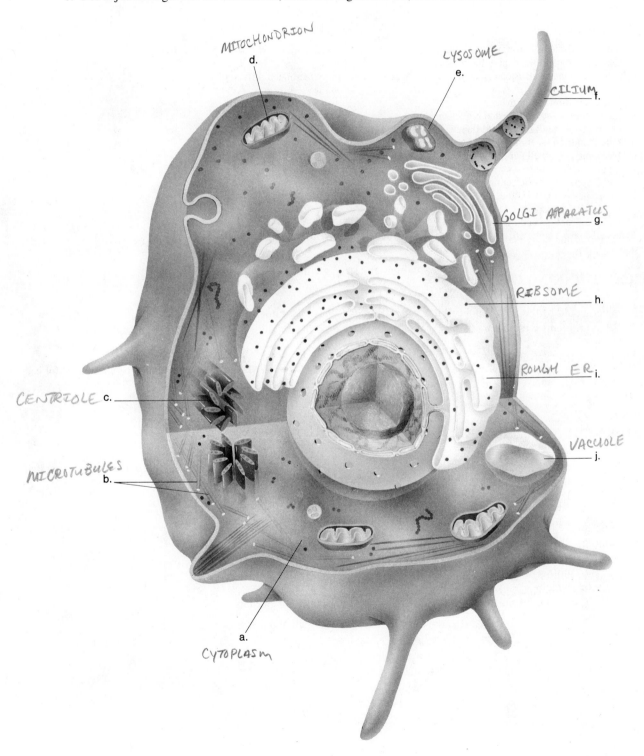

a. CYTOPLASM
b. MICROTUBULES
c. CENTRIOLE
d. MITOCHONDRION
e. LYSOSOME
f. CILIUM
g. GOLGI APPARATUS
h. RIBOSOME
i. ROUGH ER
j. VACUOLE

Cellular Comparisons (p. 37)

7. State 3 differences between animal and plant cells. PLANT CELL HAVE WALLS, HAVE CHLOROPLAST
8. Select the correct phrases pertaining to prokaryotic cells.
   a. found in bacteria
   b. contain many organelles
   c. chromosome enclosed by a nuclear membrane
   d. cells lack chloroplasts
   e. preceded eukaryotic cells

Answers:
1. a, b, e
2. a. osmosis b. osmosis c. osmosis d. active transport e. diffusion f. active transport g. diffusion h. active transport
3. a. 92 b. 0.5 c. hypotonic, hypertonic d. isotonic
4. a. both b. facilitated diffusion c. active transport d. active transport e. facilitated diffusion
5. a, b, d
6. a. cytoplasm—matrix of the cell
   b. microtubules—cell shape and movement
   c. centriole—organizes cilia and flagella
   d. mitochondrion—cellular respiration
   e. lysosome—intracellular digestion
   f. cilium—cell motion
   g. Golgi apparatus—packaging and secretion of proteins
   h. ribosome—protein synthesis
   i. rough ER—transport
   j. vacuole—storage and transport
7. Plant cells have a cell wall, have chloroplasts, and usually lack centrioles.
8. a, d, e

CRITICAL THINKING QUESTIONS

1. What would be the effect on a cell if it were to suddenly lose its mitochondria?
2. What problems with the maintenance of cell volume would plant cells experience if they lost their cell walls?

Answers:
1. The cell would be unable to extract energy from carbohydrates. The ATP harvested by this process would be unavailable for cell functions. Therefore, the cell would run down.
2. In a hypotonic environment, they would continue to accept water until they burst.

CHAPTER TEST

Indicate your answers by filling in the blanks or circling the letter. Do not refer to the text when taking this test. In 1–8, match each description to the correct cell part.

G	1. chloroplast	a.	regulates passage of substances into the cell
H	2. cytoskeleton	b.	transport channel
B	3. endoplasmic reticulum	c.	contains enzymes for digestion
C	4. lysosome	d.	site of protein synthesis
F	5. mitochondrion	e.	location of the nucleolus
E	6. nucleus	f.	site of cellular respiration
A	7. plasma membrane	g.	found in plants, not animals
D	8. ribosome	h.	maintains cell shape

In 9–14, match each description to the correct activity.

__B__ 9. active transport
__F__ 10. diffusion
__D__ 11. exocytosis
__A__ 12. facilitated transport
__C__ 13. osmosis
__E__ 14. endocytosis

a. uses carrier molecule, no energy
b. uses carrier molecule and energy
c. water enters hypertonic solution
d. large particle expelled by cell
e. large particle enters cell
f. perfume molecules spread throughout a room

Answers: 1. g 2. h 3. b 4. c 5. f 6. e 7. a 8. d 9. b 10. f 11. d 12. a 13. c 14. e

CHAPTER

Cellular Energy

CHAPTER CONCEPTS

1. Only a continual input of energy allows a cell to maintain itself and perform special functions.
2. Photosynthesis and aerobic respiration permit a flow of energy from the sun through all living things.
3. ATP is universally used as the energy currency of cells.
4. Enzymes function in metabolic pathways where each chemical reaction has its own specific enzyme.

CHAPTER REVIEW

Living organisms need a constant input of energy. Through photosynthesis, plants and algae capture light energy from the sun and convert it to the chemical energy of sugars. Animals, by eating plants and other animals, depend on this energy source.

Metabolism—the chemical reactions occurring in the cells of living things—depends on a supply of ATP and the activity of enzymes. ATP is the universal energy-carrying molecule of cells. Enzymes catalyze the steps of metabolic pathways. Many of these organic catalysts carry out their activity with helpers called coenzymes. Several conditions, such as temperature, influence enzyme activity.

By photosynthesis, light energizes the electrons of chlorophyll molecules. These energized electrons travel through either cyclic (in Photosystem I) or noncyclic (in Photosystem II) pathways of the light-dependent reactions. Oxygen is produced in these reactions. ATP from the light-dependent reactions is used to build a sugar in the light-independent reactions of photosynthesis.

The cellular breakdown of carbohydrate molecules, particularly glucose, provides energy in the form of ATP. Through glycolysis, a net of 2 ATP molecules is produced per glucose molecule. Through the additional steps of the Krebs cycle and the electron transport chain, at least 36 molecules are produced per glucose molecule. Most of this ATP is produced by chemiosmosis along the mitochondrial membrane. For most of this additional ATP production, oxygen must be available to the cell. In the absence of oxygen, pyruvate, which is the product of glycolysis, experiences fermentation.

In addition to glucose, protein, and fat molecules, other carbohydrates can generate ATP by entering various steps in the degrative pathways of glycolysis and the Krebs cycle.

VOCABULARY

The list in the first column includes some of the chapter's terms. Each term is followed by the page where it first appears. Locate each term in the chapter and read its description. Then match the meanings to the terms.

1. ATP (p. 45) G
2. chlorophyll (p. 47) d
3. enzyme (p. 46) B
4. glycolysis (p. 50) e
5. Krebs (p. 50) H
6. light-dependent (p. 48) F
7. metabolism (p. 45) c
8. PGAL (p. 50) a

a. 3-carbon molecule
b. organic catalyst
c. all chemical reactions in cells
d. light-trapping molecule
e. anaerobic process
f. a phase of photosynthesis
g. nucleotide
h. cyclic series of reactions

Answers: 1. g 2. d 3. b 4. e 5. h 6. f 7. c 8. a

LEARNING ACTIVITIES

Study the text section by section as you answer the following questions.

Enzymes (p. 46)

1. Select the correct statements about enzymes.
    a. They are globular lipid molecules.
    b. They are inorganic catalysts.
    c. They are necessary to speed up the reaction rates of cellular reactions.
    d. They are affected by pH and temperature.
2. Complete each statement with the term increases or decreases.
    a. Destroying the active site of an enzyme __DECREASES__ its activity.
    b. Generally, more substrate __INCREASES__ the rate of an enzymatic reaction.
    c. Raising the temperature over 50° C __DECREASES__ the rate of an enzymatic reaction.
    d. Lowering the pH for an enzyme that works best in highly acidic conditions __INCREASES__ the rate of an enzymatic reaction.
3. a. Label the following diagram of the chloroplast with the terms thylakoid, grana, and stroma.

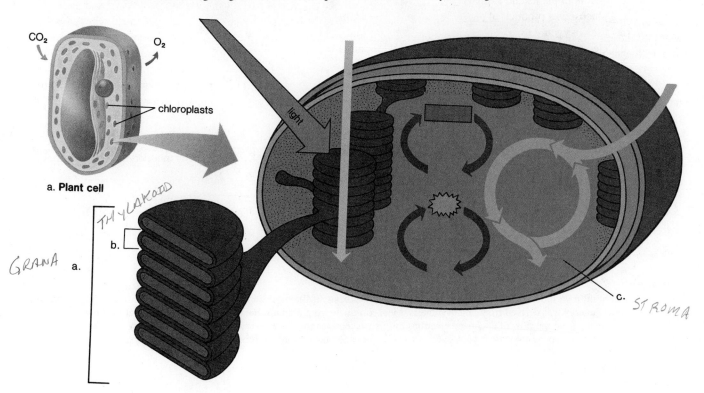

a. GRANA
b. THYLAKOID
c. STROMA

b. Which structure(s) is(are) associated with Photosystem I? with Photosystem II? A & B
c. With ATP production? AB
d. With the Calvin cycle? C

Building up Carbohydrates (p. 47)

4. Label each of the following events as indicative of light-dependent reactions (LD) or light-independent reactions (LI).
   a. ATP is produced LD
   b. Calvin cycle takes place LI
   c. chlorophyll molecules are excited LD
   d. fixation of carbon dioxide occurs LI
   e. has cyclic and noncyclic pathways LD
   f. NADPH is formed LI
   g. occurs in Photosystems I and II LD
   h. PGAL is formed LI

Breaking Down Carbohydrates (p. 50)

5. Label each of the following as describing glycolysis, the transition reaction, the Krebs cycle, or the electron transport chain.
   a. acetyl group and CoA are combined TRANSITION
   b. associated with the process of chemiosmotic phosphorylation ELECTRON TRANSPORT
   c. begins with a molecule of citric acid KREBS
   d. connects glycolysis to the Krebs cycle TRANSITION
   e. begins with glucose GLYCOLYSIS
   f. produces a net of 2 ATP molecules GLYCOLYSIS
   g. oxidation produces NADH KERBS
   h. oxygen is the final acceptor ELECTRON TRANSPORT
   i. pyruvate is oxidized and converted to an acetyl group TRANSITION
   j. produces a net of 24 ATP molecules KERBS

Comparison of Aerobic Respiration and Photosynthesis (p. 54)

6. Label each statement as describing aerobic respiration (R) or photosynthesis (P).
   a. occurs in plants only P
   b. occurs in plants and animals R
   c. glucose is a product P
   d. uses oxygen R
   e. electron transport system is in the thylakoid membrane P
   f. electron transport system is in the cristae R

Answers:
1. c, d
2. a. decreases b. increases c. decreases d. increases
3. a. A—grana, B—thylakoid, C—stroma b. A and B c. A and B d. C
4. a. LD b. LI c. LD d. LI e. LD f. LI g. LD h. LI
5. a. transition b. electron transport c. Krebs d. transition e. glycolysis f. glycolysis g. Krebs h. electron transport i. transition j. Krebs
6. a. P b. R c. P d. R e. P f. R

CRITICAL THINKING QUESTIONS

1. How is the meal that you eat today dependent on the sun?
2. How is the maintenance of life dependent on a flow of electrons in cells?

Answers:
1. Plants build food molecules by trapping energy via photosynthesis. If the meal consists of plants, such as fruits or vegetables, its chemical energy was derived from photosynthesis. Meats originate from animals that eat plants or eat plant-eating animals.
2. ATP is made from the flow of energized electrons through acceptors molecules. Without this flow, the energy to make organic carbon compounds would not exist in photosynthetic cells. These compounds are the food source of living things.

CHAPTER TEST

Indicate your answers by circling the letter. Do not refer to the text when taking this test.

1. The useful energy conversion in photosynthesis is
   a. chemical to solar.
   b. heat to mechanical.
   c. mechanical to heat.
   d. solar to chemical.
2. The presence of an enzyme _____ the rate of the appropriate chemical reaction.
   a. increases
   b. decreases
3. An enzyme, functioning best at a pH of 3, is found in a neutral environment with a temperature of 40° C. Its activity will increase by
   a. decreasing the amount of reactant.
   b. destroying the enzyme.
   c. increasing the temperature 10° more.
   d. making the pH more acidic.
4. Each is true about ATP *except* that it
   a. is a carrier of energy in cells.
   b. is common in cells.
   c. contains less energy than ADP.
   d. supplies energy to many different kinds of reactions.
5. Photosynthesis occurs best at wavelengths that are
   a. blue.
   b. gamma.
   c. infrared.
   d. ultraviolet.
6. Each is a product of the light-dependent reactions *except*
   a. ATP.
   b. NADPH.
   c. oxygen.
   d. sugar.
7. The cyclic pathways of photosynthesis produce
   a. ATP only.
   b. NADPH only.
   c. ATP and NADPH.
   d. organic sugars.
8. Fermentation is
   a. glycolysis and the Krebs cycle.
   b. glycolysis and the reduction of pyruvate.
   c. glycolysis only.
   d. the reduction of pyruvate only.
9. Each of the following is a product of aerobic respiration *except*
   a. ATP.
   b. carbon dioxide.
   c. oxygen.
   d. water.
10. Per glucose molecule, the net gain of ATP molecules from glycolysis is
    a. 2.
    b. 4.
    c. 6.
    d. 8.
11. Which is *not* an event of the transition reaction?
    a. breaks down pyruvate
    b. converts a citric acid molecule
    c. oxidized pyruvate
    d. transfers an acetyl group
12. Select the process with the greatest yield per glucose molecule.
    a. glycolysis
    b. Krebs cycle
    c. substrate level phosphorylation
    d. transition reaction

Answers: 1. d 2. a 3. d 4. c 5. a 6. d 7. a 8. b 9. c 10. a 11. b 12. b

CHAPTER

Cell Reproduction

CHAPTER CONCEPTS

1. All cells come only from preexisting cells.
2. Like begets like because the genetic material is copied and transmitted from parent to daughter cells.
3. Cell division plays a role in the cell cycle of cells and the life cycle of organisms.

CHAPTER REVIEW

Cell division accounts for growth and tissue repair in multicellular organisms. Single-celled organisms reproduce by this process.

The replication of chromosomal DNA precedes cell division. In prokaryotic cells, this chromosome is a single loop of DNA. Some cells reproduce by binary fission. Elongation of the cell, accompanied by the ingrowth of the plasma membrane, forms 2 daughter cells. Each cell receives a circular chromosome with the same genetic formation. This process occurs in prokaryotic cells and eukaryotic protists and fungi.

The number of chromosomes (diploid or haploid) in eukaryotic cells varies among species. These chromosomes are visible only when the cell is dividing. In the nondividing cell, the genetic material appears as chromatin.

Both nuclear division and cytokinesis occur during cell division in most eukaryotic cells. There are two types of nuclear division: mitosis, which produces 2 diploid cells, and meiosis, which produces 4 haploid products. In sexually reproducing organisms, mitosis accounts for growth and tissue repair; meiosis produces gametes. Mitosis is part of the cell cycle. In this cycle, interphase involves the life span of the cell when it is not dividing. Interphase has 3 phases: G_1, S, and G_2. The events of mitosis are studied over 4 stages: prophase, metaphase, anaphase, and telophase. By these events, the 2 sister chromatids of each chromosome of the dividing cell are separated and segregated into different daughter cells. Cytokinesis completes the division of the cell. By mitosis, each daughter cell receives an identical set of genetic instructions.

Meiosis involves 2 consecutive cell divisions that produce 4 haploid products. In the life cycle of animals, meiosis produces haploid sex cells through either spermatogenesis or oogenesis. The fertilization of these cells produces a diploid cell that develops into a mature organism. In plants, the haploid cells develop into a haploid, multicellular organism as part of the life cycle. Working in conjunction with fertilization and mitosis, meiosis ensures the constancy of the chromosome number and combination in a species from generation to generation.

Prior to meiosis I, the DNA of the chromosomes replicates. During meiosis I, the chromosomes of each homologous pair separate independently into different daughter cells, thus producing haploid cells. Crossing-over often occurs between nonsister chromatids in prophase of meiosis I. In meiosis II, the separation of chromosomes is similar to the pattern in mitosis. The independent assortment of chromosomes, in addition to crossing-over, account for genetic variability in the 4 daughter cells. Fertilization also accounts for variation, which establishes the potential for adaptation as a species evolves.

VOCABULARY

The list in the first column includes some of the chapter's terms. Each term is followed by the page where it first appears. Locate each term in the chapter and read its description. Then match the meanings to the terms.

1. cell cycle (p. 62)
2. centriole (p. 64)
3. centromere (p. 62)
4. chromatin (p. 62)
5. cytokinesis (p. 62)
6. fertilization (p. 71)
7. gamete (p. 66)
8. nondisjunction (p. 71)
9. synapsis (p. 67)
10. zygote (p. 66)

a. fertilized egg
b. pairing of homologous chromosomes
c. sex cell
d. union of sex cells
e. homologous chromosomes fail to separate
f. organelle organizing cell division
g. connects sister chromatids
h. tangled mass of genetic material
i. cytoplasmic division
j. has 4 phases

Answers: 1. j 2. f 3. g 4. h 5. i 6. d 7. c 8. e 9. b 10. a

LEARNING ACTIVITIES

Study the text section by section as you answer the following questions.

The Cell Cycle (p. 62)

1. State the general events of each phase of the cell cycle.
    a. M phase
    b. G_1 phase
    c. S phase
    d. G_2 phase
    e. interphase

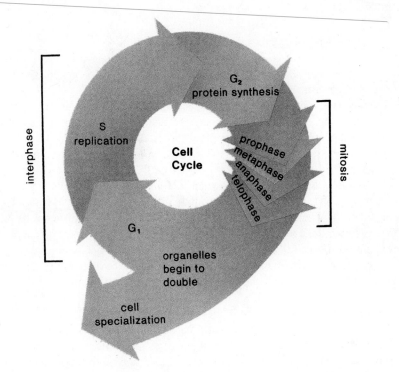

Mitosis

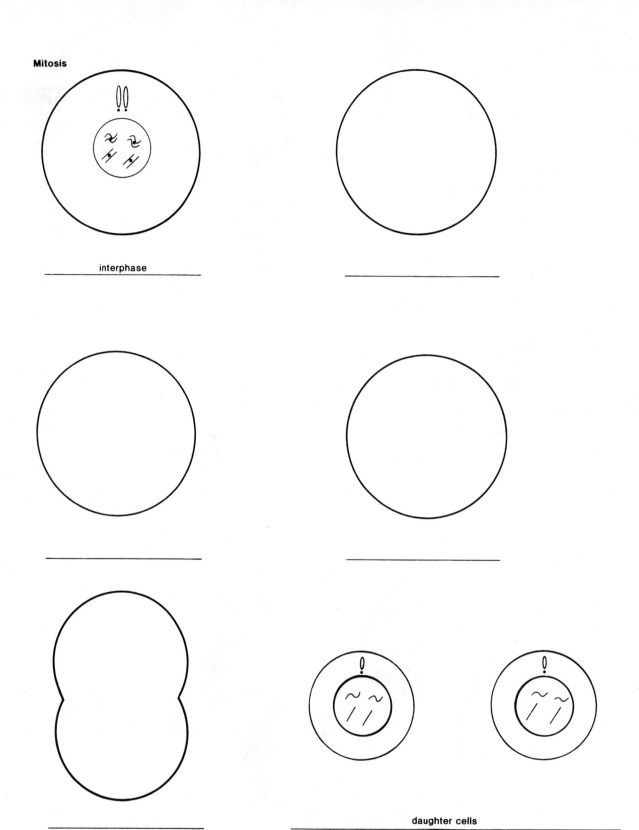

interphase

daughter cells

Mitosis and Cytokinesis (p. 62)

2. By examining the drawing of an animal cell nucleus in interphase (after replication has occurred), draw the states of mitosis. Finish your drawing with daughter cells (before replication has occurred).

3. To show the difference between plant and animal mitosis, write yes or no in the spaces provided in the table.

Mitosis	Plant Cell	Animal Cell
a. same stages		
b. spindle fibers		
c. aster		
d. cell plate		
e. furrowing		

Meiosis (p. 66)

4. Select the correct phrases pertaining to meiosis.
   a. forms gametes in animals that fuse to form a zygote
   b. forms haploid cells in the life cycle of animals
   c. produces diploid spores in plants that divide mitotically
   d. produces 4 diploid cells over 2 divisions
5. Given the following figures (p. 26) of an animal cell nucleus in interphase (after replication has occurred), fill in the blank cells with the stages of meiosis I and meiosis II.
6. Select the correct statements in the following list.
   a. Meiosis in the human male is called spermatogenesis.
   b. Oogenesis occurs in the testis.
   c. Oogenesis produces 4 functional egg cells from one cell.
   d. The human egg is larger than the human sperm.
7. In a species with 5 homologous pairs of chromosomes in diploid cells, how many chromosomal combinations can be produced by meiosis?
8. How does meiosis work with fertilization to promote variation?

Answers:
1. a. mitosis b. growth c. synthesis of DNA d. growth e. all phases but mitosis
2. see text p. 65
3. a. yes, yes b. yes, yes c. no, yes d. yes, no e. no, yes
4. a, b
5. see text p. 68
6. a, d
7. 32
8. Fertilization combines the sex cells produced meiotically in an additional variety of ways.

CRITICAL THINKING QUESTIONS

1. What is the evolutionary disadvantage of a species that can reproduce only asexually?
2. How has the evolution of cell reproduction been adaptive for multicellular organisms?

Answers:
1. Species reproducing asexually lack the potential for variation that can occur through meiosis and fertilization.
2. Cell reproduction multiplies the cell number as cells specialize during the development of the organism. Aside from this role in growth, cells die and a replacement-repair process is needed to replace these cells. Reproduction of new cells supplies that role.

Meiosis I

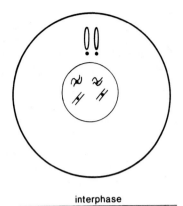
interphase

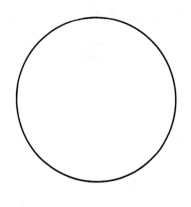

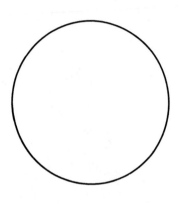

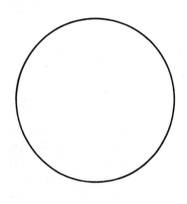

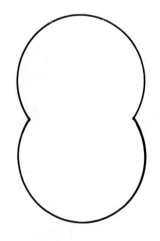

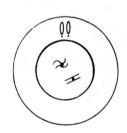
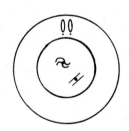
interphase

Meiosis II

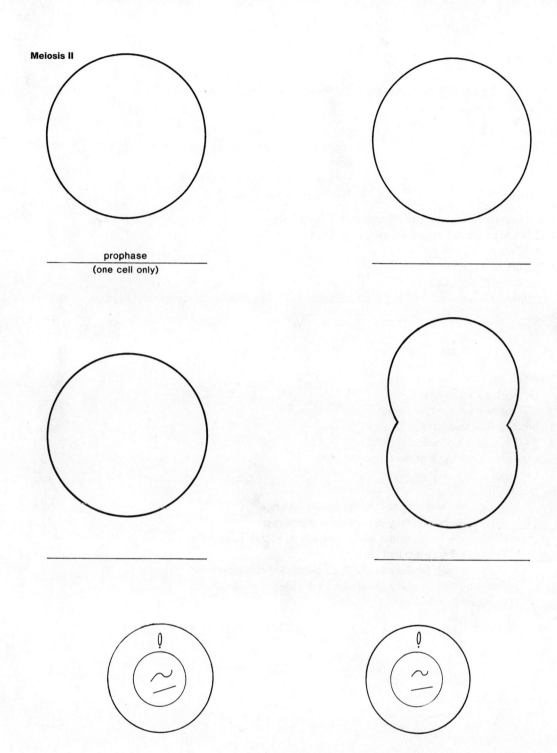

prophase
(one cell only)

daughter cells

CHAPTER TEST

Indicate your answers by circling the letter. Do not refer to the text when taking this test. In 1–5, match each description to the correct cell state.

1. anaphase c
2. interphase a
3. metaphase d
4. prophase b
5. telophase e

a. cell is not dividing
b. chromosomes first become visible
c. centromeres first split
d. chromosomes first attach to the equator
e. last stage of nuclear division

6. Select the *incorrect* statement about binary fission.
    a. DNA is replicated before division
    b. exhibits 5 active stages
    c. occurs in prokaryotic cells
    d. produces 2 daughter cells

7. The diploid chromosome number in an organism is 42. The number of chromosomes in its sex cells is normally
    a. 21.
    b. 42.
    c. 63.
    d. 84.

8. Select the *incorrect* association.
    a. G_1—cell grows in size
    b. G_2—protein synthesis occurs
    c. mitosis—nuclear division
    d. S—DNA fails to duplicate

In 9–15, match the phrase with one of the following stages.
    a. meiosis I
    b. meiosis II

9. __A__ Synapsis of homologous chromosomes occurs.
10. __A__ Separation of homologous chromosomes occurs.
11. __A__ Produces one oocyte and one polar body in human females.
12. __b__ Produces 4 sperm cells in human males.
13. __a__ Daughter cells produced have double-stranded chromosomes.
14. __b__ Daughter cells produced have single-stranded chromosomes.
15. __a__ Crossing-over occurs.

Answers: 1. c 2. a 3. d 4. b 5. e 6. b 7. a 8. d 9. a 10. a 11. a 12. b 13. a 14. b 15. a

CHAPTER

6

Mendelian Genetics

CHAPTER CONCEPTS

1. Genes, located on chromosomes, are passed from one generation to the next.
2. The Mendelian laws of genetics relate the genotype (inherited genes) to the phenotype (physical characteristics).
3. There are many exceptions to Mendel's laws, and these help explain the wide variety in patterns of gene inheritance.
4. Humans are subject to various disorders due to the inheritance of faulty genes.

CHAPTER REVIEW

Several laws of heredity explain the similarities and differences between parents and offspring over several generations. Through his breeding experiments with plants, Mendel formulated these laws. At the time Mendel started his work, the blending theory of inheritance was prevalent. Mendel disproved this theory through well-designed experiments that offered statistical evidence to the contrary.

Mendel performed several kinds of genetic crosses: the monohybrid cross, the dihybrid cross, and the testcross. From these crosses he recognized the presence of dominant and recessive factors for a trait. He also formulated the laws of segregation and independent assortment.

Since the discovery of Mendelian inheritance, several exceptions to this pattern of inheritance have been established: incomplete dominance, multiple alleles, and polygenic inheritance.

The chromosome theory of inheritance states that genes are located on chromosomes. Therefore, many patterns of gene inheritance reflect the inheritance patterns of chromosomes.

Sex determination in animals is dependent on the inheritance of chromosomes. Genes located on the X chromosome are called X-linked genes. Many X-linked traits produced by the inheritance of these genes have been discovered.

Biologists often build pedigree charts to reveal the pattern of inheritance over many generations of a family tree. Several inheritance patterns are apparent through the pedigree charts: autosomal dominant, autosomal recessive, and X-linked recessive patterns.

VOCABULARY

The list in the first column includes some of the chapter's terms. Each term is followed by the page where it first appears. Locate each term in the chapter and read its description. Then match the meanings to the terms.

1. allele (p. 79)
2. autosome (p. 85)
3. blending (p. 77)
4. genotype (p. 79)
5. incomplete dominance (p. 83)
6. monohybrid (p. 78)
7. phenotype (p. 79)
8. polygenic inheritance (p. 84)

a. genes inherited during fertilization
b. more than one gene pair acting
c. pink flowers are an example
d. alternate form of a gene
e. one trait is studied
f. the expression of the genes
g. any chromosome other than X or Y
h. incorrect theory of inheritance

Answers: 1. d 2. g 3. h 4. a 5. c 6. e 7. f 8. b

LEARNING ACTIVITIES

Study the text section by section as you answer the following questions.

Gregor Mendel/Blending Theory of Inheritance (p. 77)

1. From the following list, identify each as either the blending theory of inheritance (B) or results found in Mendel's work (M).
   a. From a cross between a red and a white flower, all of the offspring will have red flowers. _____
   b. From a cross between a red and a white flower, the offspring will have pink flowers. _____
   c. Parents of contrasting appearance will produce offspring of intermediate appearance. _____
   d. The particulate theory of inheritance occurs during crosses. _____

Monohybrid Inheritance (p. 78)

2. In peas, yellow color (Y) is dominant to green (y).

$Yy \times Yy$

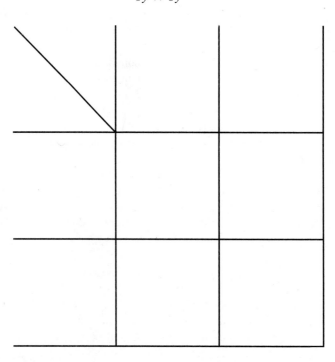

a. The genotypic ratio among the offspring from this cross is _____ YY _____ Yy _____ yy.
   The phenotypic ratio among the offspring from this cross is _____ yellow _____ green.

$$Yy \times yy$$

b. The genotypic ratio among the offspring from this cross is _____ YY _____ Yy _____ yy.
   The phenotypic ratio among the offspring from this cross is _____ yellow _____ green.

Modern Terminology and Approach (p. 79)

Complete the following statements.
3. a. The __GENOTYPE__ refers to the alleles an individual receives at fertilization.
   b. The __PHENOTYPE__ is the visible appearance of the individual's characteristics.
   c. __ALLELES__ are alternate forms of a gene.
   d. T is a symbol for a(n) __DOMINANT__ allele.
   e. t is a symbol for a(n) __RECESSIVE__ allele.

The Monohybrid Testcross (p. 79)

4. Using the letters A and a for alleles, write out the genotypes of the parents for a monohybrid testcross.
5. How many different kinds of gametes does each parent produce from this testcross?

The Dihybrid Testcross (p. 81)

6. Using the letters A, a, B, and b, write out the genotypes of the parents for a monohybrid testcross.
7. How many different kinds of gametes does each parent produce from this testcross?

Beyond Mendel's Experiments (p. 83)

8. Write the genotype of a pink snapdragon.
9. Indicate a cross between 2 pink snapdragons, summarizing the genotypic and phenotypic ratios among the offspring of the cross.
10. Write out the genotypes of a person with blood type A (heterozygous) and blood type B (heterozygous).
11. Cross the 2 genotypes from #10, summarizing the genotypic and phenotypic ratios.
12. Using 3 pairs of alleles, write the genotype of a human with the darkest skin pigmentation.
13. Using 3 pairs of alleles, write the genotype of a human with the lightest skin pigmentation.

Chromosomes and Genes (p. 84)

14. Select the correct statements pertaining to chromosomes and genes.
    a. Human females have 2 X chromosomes.
    b. Human males have 2 Y chromosomes.
    c. The genes on the X chromosomes are called sex-linked genes.
    d. Many human disorders are caused by genes on the Y chromosome.

Human Genetic Disorders (p. 86)

15. Label each human genetic disorder as autosomal dominant, autosomal recessive, or X-linked.
    a. hemophilia
    b. sickle-cell disease
    c. Tay-Sachs disease
    d. Huntington disease
    e. NF

Answers:
1. a. M b. B c. B d. M
2. a. 1/4 *YY*, 1/2 *Yy*, 1/4 *yy*; 3/4 yellow, 1/4 green
   b. 0 *YY*, 1/2 *Yy*, 1/2 *yy*; 1/2 yellow, 1/2 green
3. a. genotype b. phenotype c. Alleles d. dominant e. recessive
4. *Aa* × *aa*
5. *Aa*, 2; *aa*, 1
6. *AaBb* × *aabb*
7. *AaBb*, 4; *aabb*, 1
8. *Rr*
9. *Rr* × *Rr*
   genotype: 1/4 *RR*, 1/2 *Rr*, 1/4 *rr*
   phenotype: 1/4 red, 1/2 pink, 1/4 white
10. *IAi* × *IBi*
11. genotype: 1/4 for each of the 4 genotypes
    phenotype: 1/4 for each of the 4 phenotypes
12. *CC CC CC*
13. *cc cc cc*
14. a, c
15. a. X-linked b. autosomal recessive c. autosomal recessive d. autosomal dominant e. autosomal dominant

CRITICAL THINKING QUESTIONS

1. The phenotypic ratio of a testcross is 1:1:1:1. Was the cross producing this ratio a monohybrid testcross or a dihybrid testcross?
2. How do genes working together to determine a trait (polygenic inheritance) produce more variety among the individuals of a population?

Answers:
1. It is a dihybrid testcross. The segregation of one pair (monohybrid) of alleles from the dominant parent produces a 1:1 ratio. The independent assortment of 2 pairs of alleles from the dominant parent produces more variety, 4 kinds of alleles, and 4 different phenotypes among offspring.
2. One gene pair, with only 2 alleles, produces 3 possible genotypes. Adding more gene pairs increases the possibilities mathematically.

CHAPTER TEST

Indicate your answers by circling the letter. Do not refer to the text when taking this test.

1. When Mendel began crossing his plants, most breeders of organisms believed that
    a. dominance was complete.
    b. many genes affected one trait.
    c. red and white flowers produced pink offspring.
    d. the genetic material was always stable.

2. Two different kinds of phenotypes are produced among the offspring from a genetic cross. The genotypes of the parents are
    a. *TT* and *TT*.
    b. *TT* and *tt*.
    c. *Tt* and *tt*.
    d. *tt* and *tt*.

3. The phenotypic ratio from a genetic cross is 1:1:1:1. The genotypes of the parents are
   a. TTGG × TtGg.
   b. TtGG × Ttgg.
   c. TtGG × ttgg.
   d. TtGg × ttgg.
4. In guinea pigs the allele producing dark color (*B*) is dominant to the allele producing the light color (*b*). From a cross between 2 heterozygous organisms, the chance for producing a light-colored offspring is
   a. 4/5.
   b. 3/4.
   c. 1/2.
   d. 1/4.
5. The 2 factors of a trait separate into different sex cells during gamete formation. This statement is part of Mendel's law of
   a. dominance.
   b. independent assortment.
   c. random recombination.
   d. segregation.
6. Select the correct association.
   a. alleles—*A* or *p*
   b. heterozygous—*Aa*
   c. homozygous—*DD*
   d. homozygous—*dd*
7. From the cross *Aa* × *Aa*, the probability of producing a homozygous dominant offspring is
   a. 25%.
   b. 33%.
   c. 50%.
   d. 75%.
8. From the cross *Dd* × *Dd*, the probability of producing a homozygous dominant or heterozygous offspring is
   a. 25%.
   b. 50%.
   c. 75%.
   d. 100%.
9. How many kinds of gametes can an organism with the genotype *AaBB* produce?
   a. 1
   b. 2
   c. 3
   d. 4
10. In horses, *B* produces a black coat and *b* produces a brown coat. For another pair of alleles, *T* produces a trotter and *t* produces a pacer. A true-breeding horse, which is black and a trotter, is considered for mating. The number of kinds of sex cells it can produce is
    a. 1.
    b. 2.
    c. 4.
    d. 16.
11. How many different A-B-O blood types can be produced among offspring from the cross *IAIB* × *ii*?
    a. 1
    b. 2
    c. 3
    d. 4
12. In snapdragons, the mating of a pink plant with a white plant produces offspring that are
    a. all red.
    b. 1/2 red, 1/2 pink.
    c. 1/2 pink, 1/2 white.
    d. all white.
13. Select the human with the darkest skin pigmentation.
    a. *AaBbCc*
    b. *AaBBcc*
    c. *AaBbCC*
    d. *AabbCc*
14. Select the autosomal dominant trait.
    a. Huntington disease
    b. cystic fibrosis
    c. PKU
    d. hemophilia
15. Select the autosomal recessive trait.
    a. Huntington disease
    b. NF
    c. PKU
    d. red-green color blindness

Answers: 1. c 2. c 3. d 4. d 5. b 6. a 7. a 8. c 9. b 10. a 11. b 12. c 13. c 14. a 15. c

CHAPTER

DNA and Biotechnology

CHAPTER CONCEPTS

1. DNA is the genetic material that dictates the form, function, and behavior of organisms.
2. The genetic material is able to store information, replicate, and undergo mutation.
3. There is a flow of information from DNA to RNA to protein.
4. DNA can be manipulated and therefore we can use bacteria to produce human proteins, produce genetically engineered plants and animals, and even cure genetic disorders.

CHAPTER REVIEW

Several experiments proved that DNA is the genetic material. The work of Griffith revealed the presence of a transforming substance in the pneumococcus infecting mice. This transforming substance was DNA and not protein. The results from Hershey and Chase offered more convincing evidence for the genetic role of DNA.

Several lines of investigation contributed to the knowledge of DNA structure. For example, X-ray diffraction analysis by Franklin revealed the helical shape of the molecule. Watson and Crick used the information gained from a variety of experiments to build a model of DNA. Alternating sugar-phosphate molecules compose the sides of a ladder, with base pairs (A-T and G-C) composing the rungs. The ladder is twisted into a helix. This model also accurately predicted the mode of DNA replication. As the helix unzips, each parental strand serves as the template for the synthesis of a new daughter strand. Through semiconservative replication, each duplex produced is identical to the original double helix.

The knowledge of gene activity arose from the experiments of several investigators. Beadle and Tatum suggested the one gene-one enzyme hypothesis.

According to the central dogma, DNA serves as a pattern for its replication and also serves as a pattern to make RNA by transcription. In turn, the nucleotide sequence of messenger RNA directs the order of amino acids in a polypeptide during translation. RNA differs from DNA in several ways. The triplet code of DNA specifies the sequence of amino acids in a protein. This code is essentially universal. This code is also subject to mutation. Types include base substitutions.

Certain genes, called regulatory genes, control transcription by structural genes. Cancer is due to the disruption of genetic control in cells. Carcinogens initiate this process.

Through biotechnology, natural biological systems are manipulated to make desirable products. Plants are transgenic organisms produced through recombinant DNA technology. These plants develop desirable properties such as resistance to disease. Genetic engineering of animals is also part of current biotechnology research. Application of information gained from these studies will be used someday to treat genetic disorders.

VOCABULARY

The list in the first column includes some of the chapter's terms. Each term is followed by the page where it first appears. Locate each term in the chapter and read its description. Then match the meanings to the terms.

1. bacteriophage (p. 94)
2. codon (p. 97)
3. DNA (p. 93)
4. DNA ligase (p. 105)
5. plasmid (p. 104)
6. RNA (p. 94)
7. semiconservative (p. 95)
8. transgenic organism (p. 106)
9. translation (p. 97)
10. transcription (p. 97)

a. accessory DNA ring
b. DNA makes RNA
c. RNA makes protein
d. a virus
e. an enzyme
f. single-stranded nucleic acid
g. double-stranded nucleic acid
h. base triplet
i. type of replication
j. plant is an example

Answers: 1. d 2. h 3. g 4. e 5. a 6. f 7. i 8. j 9. c 10. b

LEARNING ACTIVITIES

Study the text section by section as you answer the following questions.

Introduction (p. 93)

1. Select the descriptions that are requirements for a substance to serve as the genetic material.
    a. able to store genetic information
    b. can be replicated
    c. can undergo mutations
    d. cannot be transmitted from generation to generation
    e. will conduct photosynthesis
2. Select the correct statements pertaining to Griffith's transformation experiments.
    a. The heat-killed S strain was not virulent.
    b. The normal R strain was virulent.
    c. The normal S strain was not virulent.
    d. A mixture of heat-killed S strain and live R strain was virulent.
    e. He worked with 2 strains of bacteria.

Structure and Function of Nucleic Acids (p. 94)

3. The following diagram shows a portion of a flattened DNA molecule and indicates its chemical composition.

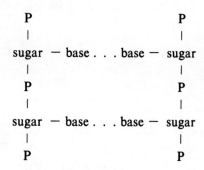

   a. Are the bases connected to the phosphate groups or to the sugar groups?

33

b. The combination of a phosphate group, a sugar group, and either a purine or pyrimidine is called a(n) _____ .
c. Why is DNA called a polynucleotide?
4. DNA is actually a long, threadlike, twisted molecule. It is composed of 2 strands that twist around each other and are connected by hydrogen bonds between the bases. Study the following figure.

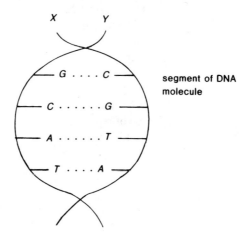

a. Which letters on the figure represent the sugar and phosphate backbone?
b. Which 4 letters or combinations of letters indicate the bases?
5. The bases of DNA are composed of 2 kinds of molecules: adenine (A) and guanine (G) are purines, and thymine (T) and cytosine (C) are pyrimidines.
a. Are 2 purines connected to form a crosspiece?
b. Are 2 pyrimidines connected to form a crosspiece?
c. What composes a crosspiece?
6. When DNA is ready to divide, or replicate, the hydrogen bonds are broken and the 2 strands come apart as if they were unzipped. The 2 separate strands may look like the following diagram.

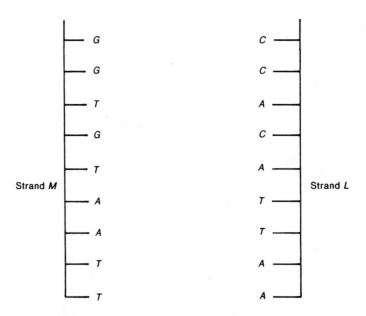

a. Draw in complementary strands to show how this diagram would appear after replication.
b. If strand M is to become a complete DNA molecule by forming a complementary half, then each
   G should attach itself to a(n) _____ .
   A should attach itself to a(n) _____ .
   C should attach itself to a(n) _____ .
   T should attach itself to a(n) _____ .

First Base	Second Base				Third Base
	U	C	A	G	
U	UUU phenylalanine	UCU serine	UAU tyrosine	UGU cysteine	U
	UUC phenylalanine	UCC serine	UAC tyrosine	UGC cysteine	C
	UUA leucine	UCA serine	UAA *stop*	UGA *stop*	A
	UUG leucine	UCG serine	UAG *stop*	UGG tryptophan	G
C	CUU leucine	CCU proline	CAU histidine	CGU arginine	U
	CUC leucine	CCC proline	CAC histidine	CGC arginine	C
	CUA leucine	CCA proline	CAA glutamine	CGA arginine	A
	CUG leucine	CCG proline	CAG glutamine	CGG arginine	G
A	AUU isoleucine	ACU threonine	AAU asparagine	AGU serine	U
	AUC isoleucine	ACC threonine	AAC asparagine	AGC serine	C
	AUA isoleucine	ACA threonine	AAA lysine	AGA arginine	A
	AUG *(start)* methionine	ACG threonine	AAG lysine	AGG arginine	G
G	GUU valine	GCU alanine	GAU aspartate	GGU glycine	U
	GUC valine	GCC alanine	GAC aspartate	GGC glycine	C
	GUA valine	GCA alanine	GAA glutamate	GGA glycine	A
	GUG valine	GCG alanine	GAG glutamate	GGG glycine	G

Protein Synthesis (p. 97)

7. If the mRNA reads UUUACAGACUGA, what is the order of amino acids directed by this sequence?
8. What is the order of mRNA codons that direct the order Glu—Trp—His of amino acids in a polypeptide?
9. If the codon for mRNA is AGACUG, what would the 2 consecutive anticodons be on tRNA?
10. Review the processes of transcription and translation by completing the following paragraph.
    DNA contains (a) _____ for protein synthesis; it is a(n) (b) _____ code because 3 bases indicate the one particular (c) _____ . During transcription, (d) _____ RNA is produced with bases that are (e) _____ to the bases in DNA. Thus, DNA serves as a(n) (f) _____ for mRNA production. While the bases in DNA are called the code, the bases in mRNA are called (g) _____ . Messenger RNA moves into the cytoplasm and becomes associated with the (h) _____ , which contain (i) _____ RNA molecules. Also in the cytoplasm are (j) _____ molecules with a(n) (k) _____ at one end and one of the 20 amino acids at the other. During translation, the tRNA molecules bring amino acids to the ribosome in the order indicated by the DNA code. The original sequence of bases in DNA orders the (l) _____ of amino acids in a protein.

Other Features of DNA (p. 101)

11. Complete the following statements.
    a. A(n) _____ gene controls the activity of a structural gene.
    b. A gene _____ is any alternation in the code of a single gene.
    c. _____ suppressor genes code for proteins that ordinarily suppress cell division.

Biotechnology (p. 102)

12. List several ways that bacteria have been genetically engineered to perform useful services.
13. From the following list, check the advantages plants possess that make them suitable for genetic manipulation.
    a. A single cell can be stimulated to grow an entire plant.
    b. Plants of agricultural significance are easily manipulated.
    c. They accept a wide variety of plasmids.
    d. They are easily grown in tissue culture.
14. List 3 ways plants have been genetically modified to improve the quality of human life.
15. Describe how gene therapy will be successful in treating human disorders.
16. Define a physical map.

Answers:
1. a, b, c
2. a, d, e
3. a. sugars b. nucleotide c. many nucleotides are joined together
4. a. X, Y b. A-T, G-C
5. a. no b. no c. purine with a pyrimidine
6. a. see text p. 95. b. C, T, G, A
7. phenylalanine—threonine—aspartate acid—stop
8. GAA or GAG—UGG—CAC or CAU
9. UCU GAC
10. a. code b. triplet c. amino acid d. messenger e. complementary f. template g. codons h. ribosomes i. ribosomal j. transfer k. anti-codon l. sequence
11. a. regulatory b. mutation c. Tumor
12. genetically engineered hormones replace hormones at deficient levels in the human body; DNA probes bind to complementary base pairs of genes of interest; genetically engineered vaccines treat human diseases
13. a, d
14. increased nutritional value to the plant; increased productivity requiring less fertilizer; increased tolerance and resistance to unfavorable environments
15. A transplanted gene will be used to treat a human disorder.
16. It represents the DNA sequence on that chromosome.

CRITICAL THINKING QUESTIONS

1. What characteristics of protein structure do you think might limit its ability to be the genetic material?
2. What do you think is the basis for a universal genetic code throughout the kingdoms of life?

Answers:
1. It does not have sufficient variety in its structure to encode genetic information.
2. All cells have a common origin. Certain characteristics of the cell have remained constant to avoid needless waste of energy and disruption of genetic systems.

CHAPTER TEST

Indicate your answers by circling the letter. Do not refer to the text when taking this test.

1. Griffith found that heat-killed S strains were
   a. mobile.
   b. not mobile.
   c. not virulent.
   d. virulent.
2. In the DNA double helix
   a. A = A.
   b. A = T.
   c. C = C.
   d. C = G.
3. Replication of DNA cannot begin until the helix
   a. joins.
   b. transcribes.
   c. transposes.
   d. unwinds.
4. Semiconservative replication means that in DNA
   a. each old strand makes one new strand.
   b. each old strand makes one old strand.
   c. 2 old strands make 2 new strands.
   d. 2 new strands make 2 new strands.

5. Select the characteristic that is *not* different between DNA and RNA.
   a. identity of the nucleotide sugar
   b. identity of one of the bases
   c. number of strands in the molecule
   d. solubility in water
6. Select the *incorrect* association.
   a. mRNA—takes DNA message to the ribosome
   b. mRNA—takes amino acids to the ribosome
   c. tRNA—carries amino acids
   d. tRNA—arrives at the ribosome
7. The base sequence of DNA is ATAGCATCC. The sequence of RNA transcribed from this strand is
   a. ATAGCATCC.
   b. CCTACGATA.
   c. CCUACGAUA.
   d. UAUCGUAGG.
8. The mRNA base sequence is UUAGCA. The 2 anticodons complementary to this are
   a. AAT CGT.
   b. AAU CGU.
   c. TTA GCA.
   d. UUA GCA.
9. The DNA base sequence changes from ATGCGG to ATGCGC. This type of mutation is
   a. deletion.
   b. frameshift.
   c. substitution.
   d. translocation.
10. A regulatory gene
    a. is the same as a structural gene.
    b. controls the activity of a structural gene.
    c. acts as a mutation.
    d. consists of protein and not DNA.
11. Select the *incorrect* description of a plasmid.
    a. used as vectors
    b. consists of chromosomal DNA
    c. found in some bacteria
    d. small, ringlike structure
12. Restriction enzymes
    a. cleave DNA into small fragments.
    b. restrict the growth of eukaryotic cells.
    c. seal pieces of DNA together.
    d. serve as introns in cells.
13. Select the human hormone that is *not* produced currently through genetic engineering.
    a. ADH
    b. growth hormone
    c. insulin
    d. tPA
14. A transgenic organism is
    a. free living and receives a foreign gene.
    b. free living and transmits a foreign gene.
    c. parasitic and receives a foreign gene.
    d. parasitic and transmits a foreign gene.
15. A physical map shows the sequence of
    a. DNA bases on a chromosome.
    b. genes on a chromosome.

Answers: 1. c 2. b 3. d 4. a 5. d 6. b 7. d 8. b 9. c 10. b 11. d 12. a 13. a 14. a 15. a

CHAPTER

Charles Darwin and the Process of Evolution

CHAPTER CONCEPTS

1. All organisms share certain common characteristics because they are descended from a common ancestor.
2. The theory of evolution explains observations from various fields of biology.
3. Evolution involves several mechanisms. Natural selection is the process by which populations accumulate adaptive traits.
4. The diversity of life is dependent upon the process of speciation. Adaptation to local environments plays an important role in speciation.

CHAPTER REVIEW

Charles Darwin's studies, in the mid-nineteenth century, led to his hypothesis of evolution. His observations of the life forms of the Galápagos Islands, including their biogeography, influenced the formation of his hypothesis.

Darwin's view of evolution supported the common descent of organisms. During this process, members of a species evolve adaptations through natural selection. Inherited variations in the members of a population establish the raw material for these adaptations. Through potential overpopulation, and the resulting competition among population members, organisms whose characteristics make them more fit are more likely to survive and reproduce. Over generations, this results in the adaptation of the population to the environment—a process called evolution.

Numerous lines of evidence currently support Darwin's theory of common descent and evolution through natural selection. Evidence includes studies from fossils, biogeography, comparative anatomy, comparative embryology, and comparative biochemistry.

Population genetics adds another dimension in understanding the mechanism of natural selection and evolution. Gene mutations produce the inheritable variations in a population.

The gene pool is the genetic makeup of a population. Mathematically, it is described through the frequency of alleles and genotypes. By the conditions stated in the Hardy-Weinberg law, these frequencies remain constant and the population does not evolve. If these conditions are not met, which is usually the case, the frequencies change and the population evolves.

Three types of selection occur in the environment: stabilizing, disruptive, and directional. There is documented evidence for each type.

Species are produced by evolution. Organisms in a species breed only among themselves and not with other species. Speciation occurs either by allopatric (i.e., adaptive radiation) or sympatric (i.e., polyploidy in plants) means. Evolution can also occur above the species level, which is a process called macroevolution.

VOCABULARY

The list in the first column includes some of the chapter's terms. Each term is followed by the page where it first appears. Locate each term in the chapter and read its description. Then match the meanings to the terms.

1. adaptation (p. 122)
2. allopatric (p. 126)
3. biogeography (p. 119)
4. fitness (p. 122)
5. Galápagos (p. 118)
6. mutation (p. 124)
7. natural selection (p. 121)
8. population (p. 123)

a. provides new alleles
b. ability of an organism to reproduce
c. mechanism of evolutionary change
d. members belong to the same species
e. islands where Darwin studied
f. type of speciation
g. trait promoting survival
h. distribution of organisms throughout the world

Answers: 1. g 2. f 3. h 4. b 5. e 6. a 7. c 8. d

LEARNING ACTIVITIES

Study the text section by section as you answer the following questions.

Charles Darwin and Evidence of Evolution (p. 119)

1. Fill in the blanks in the following paragraphs.
   Fossils: Darwin noticed a close (a) _____ between modern forms and extinct species known only through fossils. He began to think that these fossil forms might be (b) _____ to modern species. If so, the implication is that new species appear on earth as a result of biological change.
   Biogeography: Darwin noticed that whenever the environment changed, the types of species (c) _____ . He also observed that similar environments have (d) _____ adapted species. This indicates that species are suited to the environment.
   Darwin's finches: Darwin speculated that a mainland finch was the (e) _____ ancestor for all the different species of finches on the Galápagos Islands. This shows that speciation occurs.
   Conclusion: Based on his observations, Darwin came to accept the idea of (f) _____ ; that is, life forms change over time.
2. Place statements a–e in proper order, by rearranging the letters, to describe Darwin's theory of natural selection.
   a. The result of organic evolution is many different species, each adapted to specific environments.
   b. Many more individuals are produced each generation than can survive and reproduce.
   c. Gradually, over long periods of time, a population can become well adapted to a specific environment.
   d. There are inheritable variations among the members of a population.
   e. Individuals having adaptive characteristics are more likely to be selected to reproduce by the environment.
3. In each of the following pairs of situations, indicate which members are more fit.
   a. In the forest, certain ground plants
      1. are able to grow in the shade.
      2. require full sunlight.
   b. In the depths of the ocean, certain fishes
      1. need to eat only infrequently.
      2. must eat continuously.
   c. In a mountain village, some inhabitants _____ when the oxygen level falls below normal.
      1. get dizzy
      2. do not get dizzy

39

For items 4–8, match each term to the description stating how it serves as an important line of evolutionary evidence.

4. biogeography
5. comparative anatomy
6. comparative biochemistry
7. comparative embryology
8. fossil record

a. reveals the succession of life forms through preserved remnants
b. shows similarities and differences in structures among organisms
c. closely related organisms have high correlations in DNA base sequences
d. organisms arise and disperse from place of origin
e. vertebrates show a plan of unity through a study of their embryos

Population Genetics (p. 123)

9. An investigator determines that 4% of a population is albino. Answer the following questions about this population.
   a. $q^2 =$
   b. This represents the percentage of the population that is _____.
   c. What is the frequency of the recessive allele in this population? $q =$ _____.
   d. Considering the frequency of the recessive allele, what is the frequency of the dominant allele? $p =$ _____.
   e. If $p =$ this value, then $p^2 =$ _____.
   f. This is the frequency of the population that is _____.
   g. The value of $2pq =$ _____.
   h. This represents the frequency of the population that is _____.
   i. What percentage of the population has normal pigmentation?
10. Using this Punnett square, show that the next generation of the population mentioned in #9 will have exactly the same composition, assuming that evolution has not occurred.

	( )A	( )a
( )A	( )AA	( )aa
( )a	( )Aa	( )aa

   a. frequency of $A =$
      frequency of $a =$
   b. Describe the gene pool of the next generation.
   c. What does this prove?
11. Label each of the following examples as describing stabilizing selection, disruptive selection, or directional selection.
   a. Fishing regulations in a pond require keeping the bass over 5 pounds and under one pound. Weight classes in between are returned to the pond.
   b. The darkest sunfish are more likely to be captured by bass in a pond. Light-shaded sunfish blend in with the light-colored bottom of the pond and can escape predation.
   c. Trapping regulations in a pond require keeping muskrats in the 5–10-pound category. Heavier and lighter muskrats must be returned.

Speciation (p. 126)

12. What is the biological definition of a species?
13. Put the following in the proper order to describe allopatric speciation.
   a. A newly formed barrier comes between the populations.
   b. Two species now exist.

c. A species contains several interbreeding populations.
d. Divergent evolution occurs.
e. Reproductive isolation has occurred.
f. The barrier is removed.
14. Label each of the following as describing allopatric speciation or sympatric speciation.
    a. Adaptive radiation of finches on the Galápagos Islands occurred through this process.
    b. Geographical barriers prevent gene flow between variations of a species.
    c. Reproductive isolation occurs without geographical isolation.
    d. This is more common.
    e. This occurs in plants through polyploidy.

Answers:
1. a. similarity b. related c. changed d. similarly e. common f. evolution
2. d, b, e, c, a
3. a—1, b—1, c—2
4. d
5. b
6. c
7. e
8. a
9. a. 0.04 or 4% b. homozygous recessive c. 0.02 or 2% d. 0.8 or 80% e. 0.64 or 64% f. homozygous dominant g. 0.32 or 32% h. heterozygous i. 0.96 or 96%
10. a. Along each side of the Punnett square, $f(A) = 0.8$ and $f(a) = 0.2$.
    b. In the next generation, $f(AA) = 64\%$, $f(Aa) = 32\%$, and $f(aa) = 4\%$.
    c. the frequencies of the gene pool remain unchanged.
11. a. stabilizing b. directional c. disruptive
12. A species contains members that are reproductively isolated from members of another species.
13. c, a, d, f, e, b
14. a. sympatric b. allopatric c. sympatric d. allopatric e. sympatric

CRITICAL THINKING QUESTIONS

1. Albinos comprise 1% of a population. A nonscientist studies the population and claims that albinism can be removed from the population by preventing the mating of all albinos. Will this approach work?
2. Why do you think the species concept can often be difficult to test in wild populations?

Answers:
1. It will not work because 18% of the population members are heterozygotes and will protect the recessive allele as it is hidden from expression phenotypically. Albinism is caused by a recessive allele.
2. Members of a population could experience wide geographic separation. Although there is no evidence for an isolating barrier, these organisms will never meet to test whether or not they can effectively interbreed.

CHAPTER TEST

Indicate your answers (for 1–11) by circling the letter. Do not refer to the text when taking this test.

1. The Galápagos Islands are off the western coast of
   a. Africa.
   b. Asia.
   c. North America.
   d. South America.
2. Darwin claimed that there was a relationship between the beak size of finch species and their
   a. body size.
   b. flight pattern.
   c. prey.
   d. time of reproduction.
3. Select the statement that is *not* a tenet of Darwin's theory of natural selection.
   a. Members of a population have heritable variations.
   b. Members of a population will compete.
   c. Populations tend to reproduce in small numbers.
   d. Some population members have adaptive characteristics.
4. Which is *not* an example of fitness?
   a. plant has the broadest leaves
   b. plant has the greatest height
   c. predator has the keenest eyesight
   d. prey species runs the slowest

5. An adaptation promotes
   a. only the chance to reproduce.
   b. survival only.
   c. the chance to reproduce and survive.
   d. neither reproduction nor survival.
6. Vertebrate forelimbs are most likely to be studied in
   a. biogeography.
   b. comparative anatomy.
   c. comparative biochemistry.
   d. ecological physiology.
7. Which is *not* a condition of the Hardy-Weinberg law?
   a. Gene flow is absent.
   b. Genetic drift does not occur.
   c. Mutations are lacking.
   d. Random mating does not occur.
8. The criterion used to distinguish between 2 species is based on
   a. geography.
   b. physical traits.
   c. reproduction.
   d. time.
9. Which type of speciation requires a geological isolating barrier?
   a. allopatric
   b. sympatric
10. Which type of speciation occurs from polyploidy in plants?
    a. allopatric
    b. sympatric
11. Which taxonomic level is *not* related by macroevolution?
    a. genus
    b. family
    c. order
    d. species

For items 12–15, assume that 16% of the organisms in a population are homozygous recessive. Describe the current gene pool (in percentages).

12. $f(a) =$ _____
13. $f(A) =$ _____
14. $f(AA) =$ _____
15. $f(Aa) =$ _____

Answers: 1. d 2. c 3. c 4. d 5. c 6. b 7. c 8. c 9. a 10. b 11. d 12. 40% 13. 60% 14. 36% 15. 48%

CHAPTER

Origin of Life

CHAPTER CONCEPTS

1. Life is a physical and chemical phenomenon.
2. The common origin of life explains the unity of all living things.
3. The earth is always undergoing changes; today's earth is not the same as yesterday's.
4. Living things help shape the earth's physical conditions.

CHAPTER REVIEW

Life originated on earth from a climate and primitive atmosphere very different (i.e., reducing) from the one that exists today (i.e., oxygen present). Experiments by Miller simulated the energy input and overall environmental conditions that produced the first organic molecules on the ancient earth. Polymerization of these subunits led to the formation of macromolecules. The polymerized amino acids formed microspheres with properties similar to those of cells.

Prior to Miller's work, Oparin showed the possibility of coacervate droplet formation from macromolecules under defined conditions, further proving the possibility of cell formation from an organic soup.

The forerunner of modern cells, the protocell, was a heterotrophic fermenter. It became a true cell when it developed a genetic system. From the first cell evolved prokaryote cells, and eventually eukaryotic cells.

The fossil record reveals that the first true cells existed about 3.5 billion years ago. They were probably anaerobic, photosynthesizing bacteria. Other cells, with the capability of producing oxygen, evolved from them. Aerobic respiration and eukaryotic cells evolved about 1.5 billion years ago. Multicellularity and sexual reproduction evolved about 700 million years ago.

All organisms that evolved from the first cells are currently classified into five different kingdoms.

VOCABULARY

The list in the first column includes some of the chapter's terms. Each term is followed by the page where it first appears. Locate each term in the chapter and read its description. Then match the meanings to the terms.

1. eukaryote (p. 137) e
2. heterotroph (p. 135) g
3. kingdom (p. 138) b
4. microsphere (p. 134) d
5. prokaryote (p. 136) f
6. reducing (p. 133) h
7. RNA (p. 134) c
8. stromatolite (p. 136) a

    a. pillarlike structure
    b. has celllike properties
    c. macromolecule
    d. largest taxonomic category
    e. cell with organelles
    f. cell without most organelles
    g. requires preformed food
    h. atmosphere lacking oxygen

Answers: 1. e 2. g 3. b 4. d 5. f 6. h 7. c 8. a

LEARNING ACTIVITIES

Study the text section by section as you answer the following questions.

Chemical Evolution (p. 133)

1. Place the following steps in the correct order to indicate the sequence by which chemical evolution produced the first protocell.
   - 4 a. polymerization of small molecules
   - 1 b. cooling of the primitive earth
   - 5 c. metabolism involving proteins and carbohydrates
   - 6 d. production of heterotrophic fermenter
   - 2 e. energy capture affects gases
   - 3 f. production of amino acids and sugars
2. What did the experiments from each of the following scientists prove?
   a. Miller
   b. Fox
3. Why is it probable that the protocell was a heterotrophic fermenter? — PERFORMED FOOD WAS ABUNDANT IN ITS OCEAN ENVIR., BUT OXYGEN WAS NOT AVAILABLE.
4. Select the statements that are definitely correct concerning the protocell.
   a. A means of asexual reproduction was demonstrated.
   b. DNA was the genetic material.
   (c.) It carried on energy metabolism.
   (d.) It has a lipid-protein membrane.

Biological Evolution (p. 135)

5. According to the RNA hypothesis of gene formation, place the following steps in the correct order.
   - 4 a. DNA made RNA, which in turn directed protein synthesis.
   - 2 b. RNA genes conducted protein synthesis.
   - 3 c. RNA genes eventually made DNA.
   - 1 d. RNA evolved as the first genetic material.
6. How does the protein-first hypothesis differ from the RNA hypothesis?
7. a. The oldest prokaryotic fossils date from __3.5 B__ years ago. What was unique about the photosynthesis carried out by these organisms? THEY DID NOT GIVE OFF OXYGEN
   b. Cyanobacteria appeared on the earth about __2.5 B__ years ago.
   c. Eukaryotic cells evolved about __1.5 B__ years ago.
8. Label each of the following as describing either the primitive atmosphere of the earth or the current atmosphere of the earth.
   a. It exists without the ozone shield. PRIMITIVE
   b. It favors the polymerization of organic molecules. PRIMITIVE
   c. It is a reducing atmosphere. PRIMITIVE
   d. It is an oxidizing atmosphere.
   e. It tends to break down organic molecules. CURRENT
   f. Oxygen-producing autotrophs made it. CURRENT
9. Why was the evolution of multicellularity and sexual reproduction from the first eukaryotic cells important to the success of life on earth? MULTICELLULARITY ESTABLISHED THE CAPABILITY FOR THE EVOLUTION OF MORE COMPLEX ORGANISMS. SEXUAL REPRODUCTION ESTABLISHED THE GENETIC VARIABILITY FOR ADAPTATIONS.

Five Kingdoms (p. 138)

Match each kingdom to its correct description.

10. Animals e
11. Fungi d
12. Monerans a
13. Plants c
14. Protists b

a. prokaryotes
b. single-celled eukaryotes
c. photosynthetic eukaryotes
d. heterotrophic eukaryotes that are saprophytes
e. heterotrophic eukaryotes that are motile

Answers:
1. b, e, f, a, c, d
2. a. Miller supported Oparin's idea that organic molecules could be produced from the gases of the primitive atmosphere and a strong outside energy source. b. Fox showed that amino acids polymerize to form protenoids in dry heat that change into microspheres in water, with properties similar to those of cells.
3. Preformed food was abundant in its ocean environment, but oxygen was not available.
4. c, d
5. d, b, c, a
6. Proteins or polypeptides arose before DNA and RNA. As the cell evolved sophisticated energy systems, it made DNA and RNA.
7. a. 3.5 billion; they did not give off oxygen b. 2.5 billion c. 1.5 billion
8. a. primitive b. primitive c. primitive d. current e. current f. current
9. Multicellularity established the capability for the evolution of more complex organisms. Sexual reproduction established the genetic variability for adaptations.
10. e
11. d
12. a
13. c
14. b

CRITICAL THINKING QUESTIONS

1. How do the currently existing RNA viruses lend support for the origin of the first cells on the primitive earth?
2. How would current life on earth be different if sexual reproduction had not evolved?

Answers:
1. Although not living, RNA viruses do have the capacity to make DNA. There is strong evidence to suggest that the first genes in primitive cells were more likely to contain RNA.
2. Genetic variation, and the potential evolution for many types of adaptation, would be nearly absent.

CHAPTER TEST

Indicate your answers by circling the letter. Do not refer to the text when taking this test.

1. Each of the following was present in the primitive atmosphere of the earth *except*
    a. carbon dioxide.
    b. carbon monoxide.
    c. molecular nitrogen.
    d. molecular oxygen.
2. Miller's experiments produced
    a. coacervate droplets from macromolecules.
    b. inorganic substances from organic molecules.
    c. organic molecules from inorganic substances.
    d. protocells from macromolecules.
3. Select the correct sequence that occurred on the primitive earth.
    a. gases, small molecules, macromolecules, protocells
    b. macromolecules, small molecules, protocells, gases
    c. protocells, macromolecules, small molecules, gases
    d. small molecules, gases, macromolecules, protocells
4. Microspheres formed from the polymerization of
    a. amino acids.
    b. nucleotides.
    c. sugars.
    d. water.
5. Clay may have promoted the formation of macromolecules because it
    a. attracts small organic molecules.
    b. is dry.
    c. lacks zinc and iron.
    d. wards off energy.

6. Protocells definitely exhibited each of the following *except*
   a. the ability to separate from water.
   b. a lipid-protein membrane.
   c. the conduction of energy metabolism.
   d. a means of self-replication.
7. The most primitive life forms were
   a. anaerobic photosynthesizers.
   b. eukaryotic plants.
   c. eukaryotic protistans.
   d. prokaryotic cells.
8. Currently, the atmosphere's ozone
   a. enhances photosynthesis.
   b. promotes the origin of life today.
   c. protects organisms from the effect of ultraviolet rays.
   d. reacts with and destroys pollutants.
9. The kingdom of single-celled, eukaryotic organisms is the
   a. fungi.
   b. monerans.
   c. plants.
   d. protists.
10. Fungi are described as
    a. multicellular eukaryotes.
    b. one-celled eukaryotes.
    c. multicellular prokaryotes.
    d. one-celled prokaryotes.

Answers: 1. d 2. c 3. a 4. a 5. a 6. d 7. d 8. c 9. d 10. a

CHAPTER 10

Viruses and Kingdoms Monera, Protista, and Fungi

CHAPTER CONCEPTS

1. Classification of organisms reflects their evolutionary relationships.
2. The organisms living today are related to organisms that lived previously known only by the fossil record.
3. Organisms are classified from the simple to the complex, reflecting the order in which they evolved.
4. The organisms in kingdoms Fungi, Plantae, and Animalia are most likely related to organisms in the kingdom Protista.

CHAPTER REVIEW

Viruses exhibit the characteristics of life only when inhabiting host cells. As obligate parasites, they cause various diseases in plants and animals. Bacteriophages are viruses that parasitize bacterial cells.

All viruses consist of an outer protein coat and an inner core of nucleic acid. Animal viruses enter the host cell. Once inside the cell, they take over its metabolism to produce new viruses. Some animal viruses are RNA viruses.

Single-celled prokaryotes belong to the kingdom Monera. Most members are bacteria. These prokaryotes display three basic shapes. They reproduce asexually through binary fission. Some form endospores. They also differ in their tolerance and need for oxygen. Some species are autotrophic, either through photosynthesis or chemosynthesis. Most are aerobic heterotrophs, functioning as saprophytic decomposers.

Most bacteria are classified as Eubacteria. Cyanobacteria were the first organisms to carry out photosynthesis similar to the process in plants.

The protists and fungi are eukaryotes. The kingdom Protista consists of protozoans, algae, and slime molds. Protozoan cells show numerous specializations, including several types of vacuoles. Protozoans are classified according to locomotion: amoeboids, ciliates, flagellates, and sporozoans.

Algae are photosynthetic aquatic organisms. They can be single cells, filamentous, colonial, or multicellular. They are usually named by the type of pigment in their cells.

The kingdom Fungi contains organisms that are saprophytic decomposers. Their mycelium body consists of hyphae. Yeasts are fungi that are not composed of hyphae. A lichen contains both algal cells and fungal hyphae.

VOCABULARY

The list in the first column includes some of the chapter's terms. Each term is followed by the page where it first appears. Locate each term in the chapter and read its description. Then match the meanings to the terms.

1. binary fission (p. 145) F
2. endospore (p. 145) e
3. fungus (p. 142) h
4. lichen (p. 153) d
5. moneran (p. 144) c
6. protist (p. 146) b
7. spore (p. 152) G
8. virus (p. 143) A

a. requires a host to reproduce
b. eukaryote—alga
c. prokaryote
d. alga and fungus associated
e. formed in harsh environment
f. asexual reproduction
g. haploid reproductive body
h. saprophyte

Answers: 1. f 2. e 3. h 4. d 5. c 6. b 7. g 8. a

LEARNING ACTIVITIES

Study the text section by section as you answer the following questions.

Viruses (p. 143)

1. a. Label the parts of the following diagram.

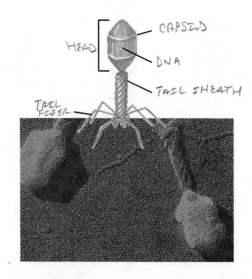

Labels: HEAD, CAPSID, DNA, TAIL SHEATH, TAIL FIBER

b. What type of virus is this? T4 BACTERIOPHAGE
c. What part enters the host? DNA
d. What is the host? BACTERIUM

2. Select the correct statements about viral reproduction.
   (a.) They must be inside a host cell to reproduce.
   b. A T-even virus attaches to a bacterial cell by its protein capsule.
   (c.) A T-even virus takes over the metabolism of a bacterial cell.
   (d.) Animal viruses can be reproducing retroviruses.
   e. For a retrovirus, DNA makes RNA.

Kingdom Monera (p. 144)

3. a. What type of cell structure do monerans exhibit? PROKARYOTIC
   b. What 2 types of organisms are in the kingdom Monera? BACTERIA, CYANOBACTERIA
   c. What do monerans lack in terms of their cell structure and mode of reproduction?
   MEMBRANOUS ORGANELLS AND ROUTINE SEXUAL REPRODUCTION.

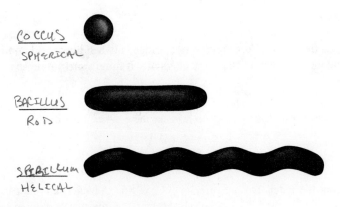

COCCUS — SPHERICAL
BACILLUS — ROD
SPIRILLUM — HELICAL

4. Label the 3 shapes of bacteria in the preceding drawing.
5. a. What is the difference between the type of nutrition carried out by chemosynthetic and autotrophic bacteria?
   b. What is the difference between the heterotrophic nutrition carried out by saprophytes and heterotrophs?
6. Describe each of the following types of respiration observed in bacteria.
   a. aerobes REQUIRE OXYGEN FOR SURVIVAL.
   b. obligate anaerobes DIE IN THE PRESENCE OF OXYGEN
   c. facultative anaerobes CAN SURVIVE EITHER IN THE PRESENCE OR ABSENCE OF OXYGEN.

Kingdom Protista (p. 146)

7. Name 3 animal-like characteristics of protozoans. USUALLY COLORLESS, MOTILE, AND CAPABLE OF ENGULFING FOOD.
8. Name and describe 3 specialized cell structures of the
   a. amoeba. FOOD VACUOLES - ENCLOSE, FOOD, PSEUDOPODIA - MOVEMENT, CONTRACTILE VACUOLE - ELIMINATIONS OF WATER.
   b. paramecium. GULLET - FOOD
9. Select the correct statements about algae.
   ⓐ They are aquatic organisms.
   b. They are commonly named by mode of reproduction.
   ⓒ *Chlamydomonas* is a member.
   ⓓ Some produce 2 strains of gametes.
   ⓔ Brown algae attach by holdfasts.
   f. Diatoms are green algae.

Kingdom Fungi (p. 152)

10. Select the correct statements about fungi.
    a. They are photosynthetic organisms.
    ⓑ Most are made up of hyphae.
    c. The spore is a diploid reproductive structure.
    ⓓ *Penicillium* is a member.
    ⓔ Yeasts are unicellular members.

Answers:
1. a. see text p. 143 b. bacteriophage c. DNA d. bacterium
2. a, c, d
3. a. prokaryotic b. bacteria and cyanobacteria c. membranous organelles and routine sexual reproduction
4. see text p. 144
5. a. Autotrophic bacteria produce food by photosynthesis, using the energy of sunlight. By chemosynthesis, the oxidation of inorganic substances generates energy for use. b. Saprophytes digest dead organic matter, whereas parasites derive nutrients from host cells.
6. a. require oxygen for survival b. die in the presence of oxygen c. can survive either in the presence or absence of oxygen
7. usually colorless, motile, and capable of engulfing food
8. a. food vacuoles—enclose food, pseudopodia—movement, contractile vacuole—elimination of water b. gullet—food intake, anal pore—elimination of wastes, macronucleus—controls cell
9. a, c, d, e
10. b, d, e

CRITICAL THINKING QUESTIONS

1. Why do some scientists think that viruses represent a bridge between living and nonliving forms?
2. How do you think that the earth would change ecologically if fungi were not present?

Answers:
1. They consist of organic substances that are not living. This combination, however, takes on a living form when inside a host cell.
2. The rate of decomposition of many substances would decrease, so these substances would not recycle as easily.

CHAPTER TEST

Indicate your answers by circling the letter. Do not refer to the text when taking this test.

1. Each is generally true of viruses *except* that they
   a. consist of a nucleic acid core.
   b. consist of a protein coat.
   c. have a specific host range.
   d. reproduce independently.
2. The first type of cells to evolve were
   a. eukaryotic.
   b. prokaryotic.
3. Select the shape that is *not* representative of some bacteria.
   a. bacillus
   b. coccus
   c. flagellar
   d. spirillum
4. Which of the following is *not* a form of sexual reproduction in bacteria?
   a. binary fission
   b. conjugation
   c. transduction
   d. transformation
5. The function of the bacterial endospore is to
   a. increase the rate of anaerobic respiration.
   b. promote asexual reproduction.
   c. protect against the attack from immune systems.
   d. withstand harsh environmental conditions.
6. A bacterium that can exist in the presence or absence of oxygen is a(n)
   a. autotroph.
   b. facultative anaerobe.
   c. obligate anaerobe.
   d. saprophyte.
7. The amoeba moves by
   a. cilia.
   b. flagella.
   c. pseudopodia.
   d. nucleii.
8. Protozoans are classified by their means of
   a. locomotion.
   b. size.
   c. reproduction.
   d. habitat preference.
9. Which is *not* true of algae?
   a. aquatic organisms
   b. producers of a community
   c. named by type of pigment
   d. protect the zygote during sexual reproduction
10. Which is *not* true of fungi?
    a. some are parasitic
    b. usually saprophytic
    c. producers of ecosystems
    d. yeasts are members

Answers: 1. d 2. b 3. c 4. a 5. d 6. b 7. c 8. a 9. d 10. c

Handwritten note (top): GAMETOPHYTE: THE HAPLOID, MULTICELLED, GAMETE-PRODUCING PHASE IN THE LIFE CYCLE OF MOST PLANTS.

Handwritten margin note: BYROPHOTES

CHAPTER 11

Plant Kingdom

Handwritten note: ALTERATION OF GENERATION: 2 GENERATION LIFE CYCLE

CHAPTER CONCEPTS

Handwritten note: CONIFERS ARE LARGEST GROUP OF GYMNOSPERMS (REDWOOD)

1. Plants have an evolutionary history.
2. Plants are adapted to living on land.
3. Plants, like other living things, have a reproductive strategy that promotes survival of the species.
4. Plants are classified according to structure and reproductive strategy.

CHAPTER REVIEW

Plants are land-dwelling organisms that carry out photosynthesis. All species have an alternation-of-generations life cycle marked by either a dominant sporophyte or gametophyte. The degree of dominance of the sporophyte is a key to the success of any kind of plant. Overall, plants have evolved a wide variety of adaptations for terrestrial success.

The bryophytes are nonvascular plants lacking specialized plant structures. Mosses have a dominant gametophyte. Fertilization produces a sporophyte, which makes spores that germinate into male and female gametophytes. This plant group is important ecologically.

Primitive members of the vascular plants are the ferns, club mosses, and horsetails. Ferns are distinguished by a water-dependent gametophyte. The sporophyte is identified by large fronds and sori.

Among the gymnosperms, the conifers are the best known. The sporophyte is dominant, with male and female cones. Pollen grains are the male gametophytes. Megaspores develop into the female gametophytes. Fertilization produces a seed. The economic importance of conifers is significant.

Angiosperms, the flowering plants, are a diverse and highly successful group. Pollination at the flower leads to a double fertilization. The presence of flowers on the plant promotes animal pollination and fruit production. Angiosperms are the main producers in the earth's ecosystems.

Handwritten notes (bottom):

VASCULAR: MEANS IT HAS TRANSPORT

VASCULAR TISSUE: ROOTS, STEMS, LEAVES. TALLEST ORGANISMS ARE VASCULAR PLANTS... IE REDWOOD TREES

STOMATA: A CONTROLLABLE GAP BETWEEN 2 GUARD CELLS IN STEMS AND LEAVES, THAT ALLOW CO_2 IN AND H_2O VAPOR OUT.

CONIFERS: ARE THE MOST TYPICALLY EX: OF GYMNOSPERM, PLANTS THAT PRODUCE NAKED SEEDS FOR DISPERSAL BY THE WIND.

ANGIOSPERS: ARE FLOWERING PLANT.

VOCABULARY

The list in the first column includes some of the chapter's terms. Each term is followed by the page where it first appears. Locate each term in the chapter and read its description. Then match the meanings to the terms.

1. angiosperm (p. 162) h
2. bryophyte (p. 158) b
3. flower (p. 162) f
4. fruit (p. 165) d
5. gymnosperm (p. 162) g
6. phloem (p. 158) e
7. pollen grain (p. 160) c
8. xylem (p. 158) a

a. transports water, minerals
b. moss is an example
c. male gametophyte
d. mature ovary
e. transports sugars
f. reproductive organ
g. has naked seeds
h. flowering plant

Answers: 1. h 2. b 3. f 4. d 5. g 6. e 7. c 8. a

LEARNING ACTIVITIES

Study the text section by section as you answer the following questions.

Introduction (p. 156)

1. Select the correct characteristics for plants.
   a. adapted to living on land
   b. diploid sporophyte produces diploid spores
   c. haploid gametophyte produces sex cells
   d. photosynthetic organisms

Nonvascular Plants (Bryophytes) (p. 158) THEY DON'T HAVE WELL-DEVELOPED ROOTS, STEMS, AND LEAVES

2. Select the correct statements about bryophytes.
   a. They are the only nonvascular plants.
   b. They lack roots, stems, and leaves.
   c. The diploid generation is dominant.
   d. The zygote develops into a gametophyte.
   e. Sphagnum is a member with commercial importance.

Vascular Plants Without Seeds (p. 158)

3. Select the correct statements about the vascular plants without seeds.
   a. Xylem and phloem are present.
   b. They contain roots, stems, and leaves.
   c. They were the dominant plants during the Mesozoic era.
   d. Ferns and club mosses are current-day members.
   e. The sporophyte generation is dominant.
   f. The gametophyte is small and heart shaped.

Seed Plants (p. 160)

4. The following figure shows the alternation-of-generations life cycle in seeds plants. Put the notation N or 2N on each line in the diagram.

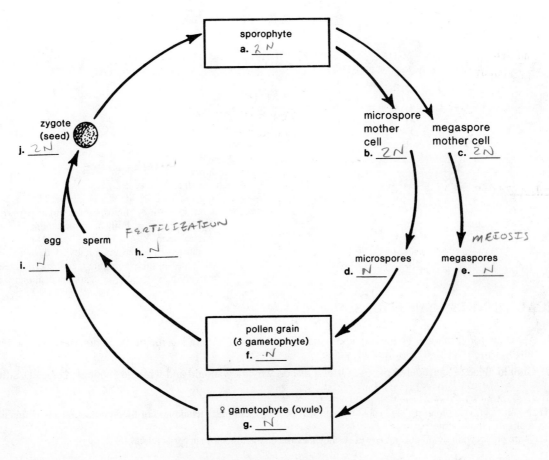

5. Label each of the following as describing an angiosperm, a gymnosperm, or both.
   a. ANGIOSPERM   Flower is the reproductive organ.
   b. ANGIOSPERM   Fruit becomes the mature ovary.
   c. ANGIOSPERM   Includes most deciduous trees of the temperate zone.
   d. " GYMNOSPERM   Megaspore develops into egg-bearing gametophyte.
   e. " "   Microspore develops into a pollen grain.
   f. ANGIOSPERM   Monocots and dicots are 2 classes.
   g. " GYMNOSPERM   Plants are heterosporous.
   h. GYMNOSPERM   Pollen grain is the male gametophyte.
   i. GYMNOSPERM   They possess needlelike leaves.
   j. GYMNOSPERM   Trees produce cones.

6. Fill in the blanks of the flowering plant life cycle.
   The (a) STAMEN of the flower have anthers that contain pollen sacs. The microspore mother cells carry out meiosis, with each microspore becoming a(n) (b) POLLEN GRAIN. The pistil contains a(n) (c) OVARY at its base. An ovary can contain one or 2 (d) OVULES. A megaspore mother cell produces (e) 4 megaspores by meiosis. One of these forms a female gametophyte, also called a(n) (f) EMBRYO SAC. As a pollen grain is transported to the stigma of the pistil, it develops a(n) (g) POLLEN TUBE that grows through the pistil. Two sperm nuclei pass through the pollen tube into the embryo sac, resulting in a(n) (h) DOUBLE fertilization. One of these sperm nuclei fertilizes to form a 3N structure, the (i) ENDOSPERM cell. The ovule develops into a (j) SEED.
   ↳ WHAT WE EAT : RICE, WHEAT.

7. Label the structures of the flower featured in this diagram.

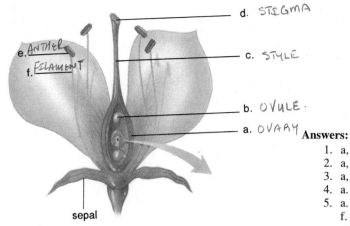

Answers:
1. a, c, d
2. a, b, e
3. a, b, d, e, f
4. a. 2N b. 2N c. 2N d. N e. N f. N g. N h. N i. N j. 2N
5. a. angiosperm b. angiosperm c. angiosperm d. both e. both f. angiosperm g. both h. gymnosperm i. gymnosperm j. gymnosperm
6. a. stamens b. pollen grain c. ovary d. ovules e. 4 f. embryo sac g. pollen tube h. double i. endosperm j. seed
7. a. ovary b. ovule c. style d. stigma e. anther f. filament

CRITICAL THINKING QUESTIONS

1. What do you think will be the impact on the life of humans if there is a major extinction of plant species during this century?
2. Should algae be considered plants? Offer a reason why they should and why they should not be classified this way.

Answers:
1. Their contribution ecologically and economically will be lost. Examples include loss of food production and maintenance of a balance of gases in the atmosphere.
2. They are photosynthetic producers; however, they lack the adaptations for success on land.

CHAPTER TEST

Indicate your answers by circling the letter. Do not refer to the text when taking this test.

1. Select the *incorrect* association.
   a. gametophyte—diploid generation
   b. gametophyte—produces sex cells
   c. sporophyte—diploid generation
   d. sporophyte—produces haploid spores
2. Select the nonvascular plant.
   a. bryophyte
   b. cycad
   c. fern
   d. rose bush
3. Spores are
   a. diploid.
   b. haploid.
4. The sporophyte is
   a. diploid.
   b. haploid.
5. Xylem and phloem are
   a. the covering tissue on roots, stems, and leaves.
   b. the male and female parts of a flower.
   c. types of flowering plants.
   d. types of vascular tissue.
6. Ferns are plants that are
   a. nonvascular with seeds.
   b. nonvascular without seeds.
   c. vascular with seeds.
   d. vascular without seeds.
7. The gametophyte is
   a. diploid.
   b. haploid.
8. Select the characteristic *not* descriptive of conifers.
   a. can withstand cold winters
   b. can withstand hot summers
   c. needlelike leaves
   d. reproduce through flowers
9. Select the *incorrect* statement about angiosperms.
   a. The gametophyte is dominant.
   b. Monocots and dicots are members.
   c. Life cycle includes double fertilization.
   d. Fruit is a mature ovary.
10. The endosperm is a source of
    a. gametes.
    b. food.
    c. shelter.
    d. transport.

Answers: 1. a 2. a 3. b 4. a 5. d 6. d 7. b 8. d 9. a 10. b

CHAPTER 12

Animal Kingdom

CHAPTER CONCEPTS

1. Animals have an evolutionary history.
2. Structural features, including embryological development, are used to classify animals.
3. Most groups of animals are marine but certain groups are adapted to living on land.
4. A jointed skeleton and segmentation occur in both arthropods and vertebrates, which are found widely on land.
5. Among mammals, primates are adapted to living in trees.
6. Humans are closely related to apes but diverged from them about 4 million years ago.

CHAPTER REVIEW

The characteristics of animals include multicellularity, heterotrophic nutrition, the ingestion of food, and usually some means of motility. Their life cycle is diploid. Most major animal phyla evolved during a short time span of the Cambrian period.

The sponges evolved independently from the other animal phyla. A tissue structure is lacking in these invertebrates. Movement is limited, although the larval stage can swim. As sessile filter feeders, the sponges exhibit several types of specialized cells, including epidermal, collar, and amoeboid cells.

Cnidarians capture their prey. They have a tissue structure, radial symmetry, and a sac body plan. Specialized structures include tentacles, a gastrovascular cavity, a nerve net, and muscle fibers. A polyp form alternates with a medusa in the life cycle.

Both parasites and free-living forms are found in the phylum of flatworms. Features in the aquatic, free-living forms include bilateral symmetry, cephalization, a ladder-type nervous system, muscles, and a gastrovascular cavity.

Mollusks have a body with a visceral mass, mantle, and foot. The coelom is reduced and the circulatory system is open. The snail and bivalves are 2 prominent members studied for specializations.

Roundworms are nonsegmented with a complete digestive tract and pseudocoelom.

Annelids are segmented worms. The coelom is well developed and the circulatory system is closed. The body also has a solid nerve cord and paired nephridia. The earthworm is a prime example but there are many marine forms.

Arthropods are the most diversified animal phylum. Jointed appendages, a chitin skeleton, cephalization, and body segmentation are major adaptations. Major classes are the crustaceans and insects.

The echinoderms have evolved radial symmetry, an internal skeleton, gills, a nerve ring, and a water vascular system. The sea star is a major echinoderm exemplifying these specializations.

Chordates have evolved a notochord, dorsal hollow nerve cord, and pharyngeal pouches at some point in their evolutionary history. Only lancelets have all of these characteristics as adults. The vertebrates develop a vertebral column in place of the notochord.

Vertebrates belong to the phylum Chordata. There are five vertebrate classes: fishes, amphibians, reptiles, birds, and mammals.

The hagfishes and lampreys are current descendants of the jawless fishes. Sharks are included among the cartilaginous fishes. Most of today's species of fishes are the ray-finned fishes.

Amphibians diversified during the Carboniferous period. They are not completely adapted to a land existence, requiring water to reproduce.

Reptiles evolved from amphibians. The evolution of a shelled egg permitted their complete adaptation for a terrestrial existence.

Adaptations of birds include feathers, constant body temperature, flight, hollow bones, and well-developed sense organs.

Mammals have evolved constant body temperature, hair, and mammary glands. Monotremes (egg-laying mammals) and marsupials evolved more unique adaptations.

Humans belong to the primates among mammals. Much is known about their evolutionary descent.

→ ANIMALS KOALAS, KANGAROOS, AND OPOSSUMS. BEGIN DEVLP. INSIDE THE FEMALE, BORN IMMATURELY. NEWBORN CRAWLS UP INTO A POUCH ON THEIRS ABDOMEN.

VOCABULARY

The list in the first column includes some of the chapter's terms. Each term is followed by the page where it first appears. Locate each term in the chapter and read its description. Then match the meanings to the terms.

1. bilateral (p. 173) J
2. coelom (p. 176) I
3. flame cell (p. 173) F
4. ganglion (p. 176) B
5. mantle (p. 176) C
6. mesoglea (p. 172) H
7. molt (p. 178) D
8. monotreme (p. 184) E
9. spicule (p. 171) G
10. vertebrae (p. 169) A

a. make up a backbone
b. nervous structure
c. contains visceral mass
d. exoskeleton is lost
e. kind of mammal
f. excretory structure
g. needle-like structure THAT SERVE AS THE INTERNAL SKELETON OF SPONGES
h. middle layer
i. body cavity
j. type of symmetry

Answers: 1. j 2. i 3. f 4. b 5. c 6. h 7. d 8. e 9. g 10. a

LEARNING ACTIVITIES

Study the text section by section as you answer the following questions.

Introduction (p. 168)

1. Select the phrases that correctly describe the characteristics of animals.
    a. autotrophic nutrition
    b. haploid life cycle
    c. ingest food
    d. carry out sexual reproduction
    e. unicellular
2. Fill in the blanks in the phylogenetic tree (p. 59).

Invertebrates (p. 169)

3. Complete the following chart by describing the function of each listed cell type.

Cell Type	Function
a. collar	
b. amoeboid	
c. epidermal	
d. spicule	
e. pore	

EVOLUTIONARY TREE OF THE ANIMAL KINGDOM.

Pg 171

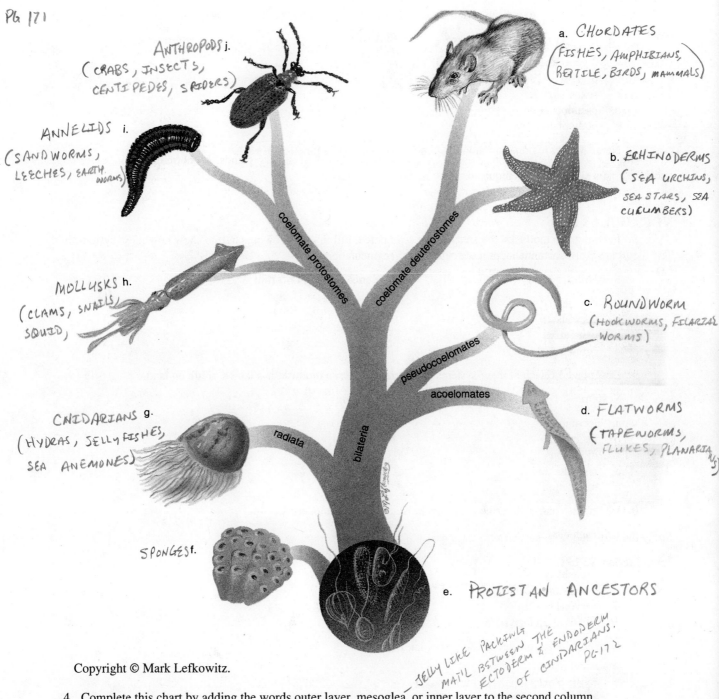

Copyright © Mark Lefkowitz.

— JELLY LIKE PACKING MATL BETWEEN THE ECTODERM & ENDODERM OF CNIDARIANS. PG-172

4. Complete this chart by adding the words outer layer, mesoglea, or inner layer to the second column.

Layers	Location
a. epidermis	OUTER LAYER
b. amoeboid cells	MESOGLEA
c. gastrodermis	INNER LAYER

172 ⅔   173

5. The free-living flatworms are more advanced than cnidarians. Indicate this by comparing hydra to planaria.

	Hydra	Planaria
a. body plan		
b. cephalization		
c. number of germ layers		
d. organs		
e. hermaphroditic		
f. symmetry		

57

6. Fill in this chart to indicate that the roundworms are more complex than the flatworms.

	Flatworms	Roundworms
a. 3 germ layers		
b. organs		
c. sexes separate		
d. pseudocoelom		
e. body plan		

7. Name 3 body regions common to all mollusks, and give a function of each.

8. In contrast to mollusks, the annelids are segmented. Fill in the following chart to show how these systems in the earthworm, an annelid, provide evidence of segmentation.

System	Evidence of Segmentation
a. circulatory	
b. nervous	
c. excretory	
d. coelom	

9. Describe how each of these systems in the grasshopper, a mollusk, is adapted to life on land.

System	Adaptations
a. locomotion	
b. excretion	
c. digestion	
d. reproduction	
e. respiration	

10. Fill in this chart to describe the characteristics of the echinoderms.

Characteristics	Description
a. type of symmetry	
b. skeletal system	
c. respiratory system	
d. nervous system	
e. water vascular system	

11. List the 3 basic characteristics of chordates.

Vertebrates (p. 181)

12. List the 6 characteristics distinguishing the vertebrates.

13. Label each statement as unique for the fishes, amphibians, reptiles, birds, or mammals.
    a. _____ Diversified during Carboniferous period.
    b. _____ Earliest vertebrate fossils.
    c. _____ First to evolve egg with extraembryonic membranes.
    d. _____ Have a swim bladder.
    e. _____ Have body hair for temperature regulation.
    f. _____ Have hollow, very light bones.
    g. _____ Monotremes and marsupials.
    h. _____ Only modern mammals with feathers.
    i. _____ Skin is a respiratory organ.
    j. _____ 3-chambered heart.

Human Evolution (p. 185)

14. Select the correct statements about human evolution and characteristics.
    a. Bipedalism distinguishes humans from the apes.
    b. *Homo habilis* dates from about 1 million years ago.
    c. *Homo erectus* dates from about 3 million years ago.
    d. Hominid fossils reveal much about human evolution.
    e. Humans are anthropoids.
    f. Humans are more closely related to monkeys than apes.
    g. Primates have mobile limbs, acute vision, and care for offspring.
    h. The first primates were prosimians.

Answers:
1. c, d
2. a. chordates b. echinoderms c. roundworms d. flatworms e. protistan ancestors f. sponges g. cnidarians h. mollusks i. annelids j. arthropods
3. a. produces water currents, captures food b. distributes nutrients and produces gametes c. protection d. internal skeleton e. entrance of water
4. a. outer layer b. mesoglea c. inner layer
5. a. sac, sac b. no, yes c. 2, 3 d. no, yes e. no, yes f. radial, bilateral
6. a. yes, yes b. yes, yes c. no, yes d. no, yes e. sac, tube within a tube
7. visceral mass—internal organs, mantle—carries visceral mass, muscular foot—locomotion
8. a. branches in every segment b. ganglia in every segment c. nephridia in every segment d. partitioned between segments
9. a. wings and legs b. excretes solid wastes c. grinding mouth parts d. fertilization and zygote production e. tracheae conduct air
10. a. radial b. internal c. gills d. nerve ring with arm extensions e. used for locomotion
11. dorsal hollow nerve chord, notochord, pharyngeal pouches
12. 3 chordate characteristics, jointed internal skeleton, extreme cephalization, paired limbs, closed circulatory system, efficient respiration
13. a. amphibians b. fishes c. reptiles d. fishes e. mammals f. birds g. mammals h. birds i. amphibians j. reptiles
14. c, d, e, g, h

CRITICAL THINKING QUESTIONS

1. From sponges to roundworms, evolution has produced a more complex body form. What evidence supports this?
2. The arthropods are considered the most successful animal phylum inhabiting the earth. What justifies this claim?

Answers:
1. Among roundworms, the body plan has reached the complexity of a tube within a tube, with bilateral symmetry and an organ level of organization. Also, a pseudocoelom is present.
2. They are the most diversified phylum, with a wide variety of species filling numerous ecological niches.

CHAPTER TEST

Indicate your answers by circling the letter. Do not refer to the text when taking this test.

1. Select the *incorrect* association.
    a. amoeboid cell—digestion
    b. collar cell—water current
    c. epidermal cell—covering
    d. spicule cell—reproduction
2. The gastrovascular cavity functions for
    a. digestion only.
    b. transport only.
    c. both digestion and transport.
    d. neither digestion nor transport.
3. Which of the following is *not* an embryonic germ layer among cnidarians?
    a. ectoderm
    b. endoderm
    c. mesoderm
4. Cephalization means that
    a. a definite head develops.
    b. reproduction is sexual.
    c. the nervous system is ladderlike.
    d. wastes are excreted through flame cells.
5. Select the *incorrect* association.
    a. cnidarian—bilateral symmetry
    b. flatworm—bilateral symmetry
    c. sponge—spicule
    d. sponge—tissues
6. In an open circulatory system
    a. blood is not always contained in vessels.
    b. blood leaves the body.
    c. the heart does not work with the lungs.
    d. the heart has 2 chambers.
7. Select the statement *not* descriptive of annelids.
    a. excretion by nephridia
    b. sac body plan
    c. segmented body
    d. well-developed coelom
8. The molting by arthropods means that they
    a. circulate blood through a closed system.
    b. move through jointed appendages.
    c. reproduce sexually.
    d. shed their exoskeletons.
9. Select the *incorrect* association.
    a. echinoderm—invertebrates
    b. echinoderm—tube feet
    c. mollusks—mantle
    d. mollusks—jointed appendages
10. Select the characteristic that does *not* belong to chordates.
    a. dorsal hollow nerve cord
    b. dorsal supporting rod
    c. gill slits absent in the adult
    d. notochord absent
11. Each is a vertebrate characteristic *except*
    a. bilateral symmetry.
    b. coelom development.
    c. open circulatory system.
    d. segmentation.
12. The skin of amphibians functions mainly for
    a. circulation.
    b. excretion.
    c. reproduction.
    d. respiration.
13. The extraembryonic membrane in the reptile promotes
    a. reinforcement against drying out.
    b. complete independence from the water for reproduction.
    c. enhanced elimination of wastes from the embryo.
    d. increased hardness to prevent breakage.
14. Feathers of birds are modified
    a. fish fins.
    b. mammalian hair.
    c. reptilian scales.
    d. vertebrate teeth.
15. Hair of mammals is an adaptation for
    a. camouflage in all species.
    b. control of body temperature.
    c. faster locomotion.
    d. regulation of waste elimination.

Answers: 1. d 2. c 3. c 4. a 5. a 6. a 7. b 8. d 9. d 10. d 11. c 12. d 13. b 14. c 15. b

CHAPTER 13

Plant Organization and Growth

CHAPTER CONCEPTS

1. Plants, like other living things, are highly organized.
2. The structure of plant organs suits their function.
3. Flowering plants are adapted to living on land.
4. The roots, stems, and leaves of plants are modified in specific ways.

CHAPTER REVIEW

The body of a flowering plant consists of a root system and a shoot system. Meristem is embryonic tissue. Three types of mature tissue are found in each: dermal, ground, and vascular. Epidermis is the dermal tissue in most parts of the plant. Parenchyma and sclerenchyma are types of ground tissue. Xylem, transporting water and minerals, and phloem, transporting organic solutes, are the types of vascular tissue.

The roots of the root system anchor the plant and absorb water. Primary growth causes the root to grow longer. The layers of a root, revealed by a cross section, are the epidermis, cortex, endodermis, and vascular cylinder.

The stem of the shoot system provides mechanical support and transports substances from the roots to the leaves. Herbaceous stems experience only primary growth. The stems of woody, temperate plants experience pronounced secondary growth, developing annual rings of xylem.

The leaves of the shoot system are the photosynthetic organs of the plant. The venation differs between monocots and dicots, the 2 classes of flowering plants. Specializations of the leaf include the epidermis, stomata, and mesophyll cells.

Macronutrients and micronutrients are necessary for the health and successful growth of plants.

VOCABULARY

The list in the first column includes some of the chapter's terms. Each term is followed by the page where it first appears. Locate each term in the chapter and read its description. Then match the meanings to the terms.

1. cotyledon (p. 199)
2. endodermis (p. 200)
3. epidermis (p. 199)
4. macronutrient (p. 204)
5. meristem (p. 198)
6. micronutrient (p. 204)
7. parenchyma (p. 198)
8. xylem (p. 199)

a. vascular tissue
b. Mn is an example
c. covering tissue
d. embryonic tissue
e. young leaf
f. ground tissue
g. contains Casparian strip
h. K is an example

Answers: 1. e 2. g 3. c 4. h 5. d 6. b 7. f 8. a

LEARNING ACTIVITIES

Study the text section by section as you answer the following questions.

Organization of a Plant Body (p. 198)

1. Label each statement as describing the root system, shoot system, or both.
   a. absorbs water
   b. anchors the plant in the soil
   c. hold the leaves
   d. stores starch
   e. transports water nutrients
2. Label each of the following statements as true or false.

   Dermal Tissue:
   a. Epidermis covers the entire body of nonwoody plants.
   b. Epidermis covers the entire body of young woody plants.
   c. Epidermis protects inner body parts.
   d. Epidermis prevents the plant from drying out.

   Ground Tissue:
   e. Parenchyma is a specialized tissue.
   f. Sclerenchyma consists of living cells.
   g. Sclerenchyma offers support to the plant.

   Vascular Tissue:
   h. Phloem transports organic nutrients from leaves to roots.
   i. Sieve-tube cells are found in xylem.
   j. Tracheids are a type of cell in phloem.
   k. Xylem transports water from roots to leaves.
3. Label each of the following statements as describing a monocot or dicot.
   a. vascular bundles scattered in stem
   b. leaf veins form a net pattern
   c. flower parts in fours or fives or multiples of four and five
   d. vascular bundles in a distinct ring

The Root System (p. 199)

4. Complete this table by naming and describing the regions beyond the root cap seen in a longitudinal section of a root.

Name of Region	Definition
a. root cap	
b.	
c.	
d.	

5. Label the drawing of a root cross section (p. 65) using these terms: epidermis, cortex, endodermis, pericycle, xylem, cambium, phloem. Define each term in the blanks provided.
   a.
   b.
   c.
   d.
   e.
   f.
   g.
   h.

The Shoot System (p. 202)

6. How can you identify one season's growth of a woody twig?
7. How can you identify one season's growth when observing the cross section of a tree trunk?

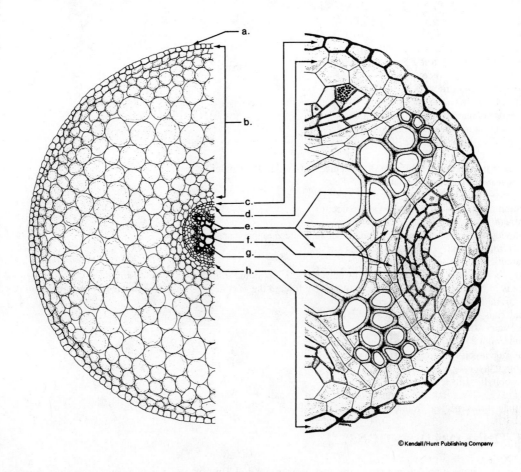

8. Complete this table to describe the anatomy of a woody stem.

Portion	Tissue Present	Function
bark	cork	b.
	phloem	c.
wood	a.	d.
pith	pith	e.

9. How does the arrangement of vascular tissue of the herbaceous dicot stem differ from that of the woody stem?

Leaves (p. 206)

10. Label this diagram of a leaf and define the portions listed.

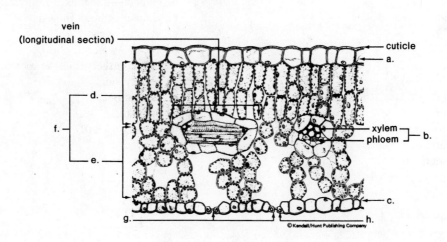

Answers:
1. a. root b. root c. shoot d. root e. both
2. a. true b. true c. true d. true e. false f. false g. true h. true i. false j. false k. true
3. a. monocot b. dicot c. dicot d. dicot
4. a. protective cap of root
   b. zone of cell division—meristem constantly producing new cells
   c. zone of elongation—cells become longer
   d. zone of maturation—cells become specialized
5. a. epidermis—outer layer of cells
   b. cortex—storage
   c. endodermis—regulates entrance of substances into vascular cylinder
   d. pericycle—produces secondary roots
   e. xylem—transports water
   f. cambium—produces new xylem and phloem
   g. phloem—transports organic nutrients
   h. endodermis—regulates entrance and exit of substances to the vascular cylinder
6. extends from one bud scale scar to the next
7. each annual ring is a season's growth
8. a. xylem b. protection c. transports organic nutrients d. support and transport of water e. storage
9. Xylem is in bundles toward the inside with phloem toward the outside. Although not situated in rings, the bundles do separate the cortex from the central pith.
10. a. epidermis—outer layer of the leaf
    b. leaf vein—transport
    c. epidermis
    d. palisade layer—photosynthesis
    e. spongy layer—gas exchange
    f. mesophyll—fills leaf interior
    g. guard cells—regulate the opening and closing of stomata
    h. cuticle—prevents leaf from drying out

CRITICAL THINKING QUESTIONS

1. Do plants have a true organ structure, just as animals do? Support your answer by describing characteristics of their anatomy.
2. Why is the evolution of xylem and phloem such an important adaptation for land success of plants?

Answers:
1. Yes. Several specialized tissues are integrated structurally in the leaf, stem, and root. All are organs, consisting of 2 or more tissues working together.
2. These types of vascular tissue allow the plant to transport substances over long distances within the plant body, from roots to leaves and vice versa.

CHAPTER TEST

Indicate your answers by circling the letter. Do not refer to the text when taking this test.

1. Each is a mature tissue in plants *except*
   a. dermal.
   b. ground.
   c. meristem.
   d. vascular.
2. Select the *incorrect* association.
   a. K—macronutrient
   b. Ca—macronutrient
   c. Cl—micronutrient
   d. N—micronutrient
3. Select the correct association.
   a. phloem—minerals
   b. phloem—photosynthesis
   c. xylem—sugar
   d. xylem—water
4. The zone farthest from the root cap is the zone of
   a. cell division.
   b. elongation.
   c. maturation.
   d. primary growth.

5. The cortex of the root functions for
   a. entrance of substances into the root.
   b. photosynthesis.
   c. protection.
   d. starch storage.
6. In the vascular arrangement of the woody stem
   a. phloem develops toward the outside.
   b. xylem develops toward the inside.
   c. Both a and b are true.
   d. Neither a nor b is true.
7. A node is the point on a stem where leaves or buds are attached.
   a. true
   b. false
8. Herbaceous stems exhibit only primary growth.
   a. true
   b. false
9. Cork cambium is located in the vascular cylinder.
   a. true
   b. false
10. Wood in the stem is converted phloem.
    a. true
    b. false

Answers: 1. c 2. d 3. d 4. c 5. d 6. c 7. a 8. a 9. b 10. b

CHAPTER 14

Plant Physiology and Reproduction

CHAPTER CONCEPTS

1. Xylem and phloem transport in plants is dependent upon the chemical and physical properties of water.
2. Plants respond to outside stimuli by changes in their growth patterns.
3. Plants have hormones that promote growth and those that inhibit growth.
4. The flowering plant life cycle is in tune with the seasons.

CHAPTER REVIEW

Water passes from the soil through the various layers of the root, eventually entering root cells by an osmotic pressure. From the root, water ascends through the xylem by the cohesion-tension principle between water molecules. It is lost through the stomata of leaves. Guard cells regulate the opening and closing of stomata in response to various environmental conditions.

Organic solutes pass through the phloem of the plant from a source to a sink, according to the pressure-flow model.

Hormones provide the signals that control the growth patterns in plants. Auxins promote cell elongation, accounting for plant phototropisms. Gibberellins also promote cell elongation. Cytokinins cause cell division and prevent leaf senescence. Ethylene causes fruit ripening. ABA is the stress hormone.

Tropisms are responses by plants to external stimuli. Both positive and negative phototropisms result from auxin migration in the plant.

Photoperiodism is the response to the relative length of daylight and darkness. Flowering in plants is a photoperiodic event, leading to the classification of short-day or long-day plants. The response of the pigment phytochrome is involved in the flowering event.

Angiosperms reproduce sexually by an alternation-of-generations life cycle. The diploid sporophyte is the dominant generation. The flower is the reproductive organ of an angiosperm. It exhibits many specialized parts: the outer whorl, inner whorl, pistil, and ovary. Double fertilization occurs in angiosperms, producing a 2N zygote and 3N endosperm cell. The zygote becomes the sporophyte embryo. The ovule of the ovary matures into a seed.

VOCABULARY

The list in the first column includes some of the chapter's terms. Each term is followed by the page where it first appears. Locate each term in the chapter and read its description. Then match the meanings to the terms.

1. ABA (p. 218)
2. auxin (p. 216)
3. cytokinin (p. 217)
4. photoperiodism (p. 211)
5. phototropism (p. 218)
6. phytochrome (p. 221)
7. pistil (p. 222)
8. sieve-tube (p. 214)
9. tracheid (p. 211)
10. transpiration (p. 212)

a. loss of water
b. phloem cell
c. pigment
d. flower part
e. IAA
f. stress hormone
g. affects cell division
h. xylem cell
i. movement toward light
j. length of darkness, light

Answers: 1. f 2. e 3. g 4. j 5. i 6. c 7. d 8. b 9. h 10. a

LEARNING ACTIVITIES

Study the text section by section as you answer the following questions.

Water Transport (p. 211)

1. Write these root structures in the order (from first to last) that water encounters them as it enters the vascular cylinder: Casparian strip, cortex, endodermis, epidermis, xylem.

2. Answer the following questions about the cohesion-tension model of transport.
    a. What 2 types of conducting cells are found in xylem?
    b. Which of these cell types forms a continuous and completely hollow pipeline?
    c. Why is cohesion necessary for the ascent of sap through xylem?
    d. What creates the tension as sap ascends through xylem?

3. Place these phrases about the opening and closing of stomata in the correct order, using letters a–e.

    _____ stomata closed
    _____ water enters guard cells
    _____ stomata open
    _____ K^+ enters guard cells
    _____ carbon dioxide decreases in leaf

Organic Nutrient Transport (p. 214)

4. Label each of the following statements as true or false.
    a. The movements of substances through phloem is called transpiration.
    b. Companion cells are not living in a mature form.
    c. The pressure-flow model accounts for the transport through phloem.
    d. Substances move in the phloem from the sink to the source.

Plant Responses to Environmental Stimuli (p. 215)

5. Complete the following table on plant hormones.

Hormone	Function
auxins	a.
b.	stem elongation
c.	cell division
ABA	d.
ethylene	e.

67

6. Describe each of these types of tropisms, indicating the hormones responsible.
   a. phototropism
   b. gravitropism
7. Study the following diagram:

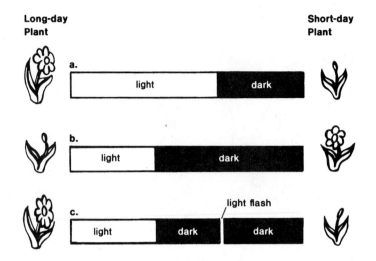

   a. A long-day plant flowers when the night is _____ than a critical length. A short-day plant will not flower when the night is _____ than a critical period.
   b. A long-day plant will not flower when the night is _____ than a critical length. A short-day plant will flower when the night is _____ than a critical length.
   c. A long-day plant will flower if a night that is _____ than a critical length is interrupted by a flash of light. A short-day plant will not flower if a night that is _____ than a critical length is interrupted by a flash of light.

Reproduction in Flowering Plants (p. 221)

8. Select the correct statements about flowering plants.
   a. The sepals form the green whorl of the flower.
   b. The petals form the colored whorl of the flower.
   c. The stigma, style, and ovary are parts of the pistil.
   d. The gametophyte is diploid.
   e. The sporophyte is haploid.
   f. A double fertilization occurs in the embryo sac.
   g. The endosperm is a food source.
   h. The cotyledons are the number of developing roots in the embryo.

Answers:
1. epidermis (first), cortex, Casparian strip, endodermis, xylem (last)
2. a. vessels and tracheids b. tracheids c. the water is pulled along because the water molecules are attracted to each other. d. transpiration
3. a. stomata close b. carbon dioxide decreases in leaf c. K^+ enters guard cells d. water enters guard cells e. stomata open
4. a. false b. false c. true d. false
5. a. cell elongation b. gibberellins c. cytokinins d. dormancy e. fruit ripening
6. a. positive is the bending of the plant stem toward a light source—auxins b. positive is growth toward the earth's center of gravity—ABA
7. a. shorter, shorter b. longer, longer c. longer, longer
8. a, b, c, f, g

CRITICAL THINKING QUESTIONS

1. Physical and chemical laws explain many of the phenomena in biological systems. How is this demonstrated in the ascent of sap through xylem?
2. Animal hormones are known to coordinate a wide variety of responses, ranging from growth and water balance to mineral balance and carbohydrate metabolism. Is this variety of response as great in plants?

Answers:
1. The evaporation of water and the cohesion between water molecules bring about this process.
2. No, not nearly as varied. It is confined mainly to growth and flowering responses.

CHAPTER TEST

Indicate your answers by circling the letter. Do not refer to the text when taking this test.

1. Water moves through the layers of a root by osmosis.
   a. true
   b. false
2. The tracheids of xylem are actively metabolizing cells.
   a. true
   b. false
3. The activity of guard cells regulates the opening and closing of stomata.
   a. true
   b. false
4. The cortex is the second root layer encountered by water as it passes through the root toward the vascular cylinder.
   a. true
   b. false
5. By the pressure-flow model of phloem transport, substances travel from the source to the sink.
   a. true
   b. false
6. The main effect of gibberellins is to
   a. hasten the ripening of fruits.
   b. inhibit the flowering process.
   c. prevent leaf abscission.
   d. promote cell division and enlargement.
7. Senescence is the
   a. aging of the overall plant.
   b. loss of leaf color.
   c. propagation of cuttings.
   d. process of flowering.
8. Another name for ABA is the
   a. dormancy hormone.
   b. flowering inhibitor.
   c. growth inhibitor.
   d. stress hormone.
9. Ovules are found in the
   a. ovary.
   b. stigma.
   c. style.
   d. petal.
10. Which seed part becomes a portion of the stem?
    a. epicotyl
    b. hypocotyl

Answers: 1. a 2. b 3. a 4. a 5. a 6. d 7. a 8. d 9. a 10. b

CHAPTER

Animal Organization and Homeostasis

CHAPTER CONCEPTS

1. The body of an animal has levels of organization.
2. All organisms carry out life processes. In animals, organ systems often carry out these functions.
3. The internal environment of organisms stays relatively constant.

CHAPTER REVIEW

As one step in the levels of organization of the animal body, tissues exist in four major categories. Epithelial tissue covers the body and lines body cavities. Its cells exist in several shapes, and it can be either simple or stratified. Connective tissue exists in many forms: loose, fibrous, adipose, cartilage, and blood. Each type has a specialized function. Muscle tissue contracts. Its forms are skeletal, cardiac, and smooth. Nervous tissue, consisting of neurons, sends signals. Glial cells support and protect neurons.

Tissues work together as organs. Many of these organs in the human body are located in the dorsal or ventral body cavities. Several related organs work in an organ system.

Through negative feedback, organs respond homeostatically to maintain a relatively constant internal environment. Regulation of body temperature is a good example of a homeostatic mechanism.

VOCABULARY

The list in the first column includes some of the chapter's terms. Each term is followed by the page where it first appears. Locate each term in the chapter and read its description. Then match the meanings to the terms.

1. adipose (p. 231)
2. appendicular (p. 233)
3. exocrine (p. 231)
4. ligament (p. 231)
5. matrix (p. 231)
6. neuron (p. 235)
7. plasma (p. 249)
8. smooth (p. 234)
9. squamous (p. 230)
10. tendon (p. 231)

a. connects muscle to bone
b. between bone cells
c. type of muscle tissue
d. gland with a duct
e. branch of the skeleton
f. liquid portion of blood
g. fat tissue
h. nerve cell
i. connects bone to bone
j. refers to flat cell

Answers: 1. g 2. e 3. d 4. i 5. b 6. h 7. f 8. c 9. j 10. a

LEARNING ACTIVITIES

Study the text section by section as you answer the following questions.

Types of Tissues (p. 230)

1. Name the 4 main tissue types and state a function of each.
2. Select the correct statements about epithelium.
    a. Ciliated epithelium lines the respiratory tract.
    b. Epithelial cells cover the body surfaces.
    c. Exocrine and endocrine glands consist of modified epithelial tissue.
    d. Pseudostratified consists of many layers.
    e. Squamous, cuboidal, and columnar epithelium are tissue types based on the number of layers.
3. Complete the following table on connective tissue types.

Connective Tissues	Locations	Functions
a. loose		
b. adipose		
c. fibrous		
d. cartilage		
e. bone		
f. blood		

4. Label each of the descriptions of muscle tissue as smooth or striated.
    a. displays light and dark bands
    b. found in the heart
    c. found in the stomach
    d. make up skeletal muscles
    e. involuntary only
5. Complete the following.
    a. A neuron sends a signal from its _____ to its _____ and on through its _____ .
    b. The function of a neuron is to _____ .
    c. Nervous tissue consists of neurons and _____ cells.
    d. This latter type of cell provides _____ to neurons and keeps tissue free of debris.
6. Name the system that performs each of the following functions.
    a. transports materials by the blood _____
    b. breaks down substances in the diet _____
    c. distribution and exchange of gases _____
    d. pull on bones to produce movements _____
    e. controls and coordinates through hormones _____
7. Label the cavity—dorsal or ventral—where each of the following organs is found.
    a. brain
    b. heart
    c. spinal cord
    d. stomach
    e. urinary bladder

Homeostasis (p. 236)

8. Define homeostasis.
9. Select the correct statements about negative feedback control.
    a. A stimulus is met by a response.
    b. Body temperature is controlled by this process.
    c. The control of blood pressure is an example.
    d. The effect of the stimulus is increased.

10. Complete the following paragraph dealing with body temperature control.
    If body temperature falls below normal, (a) _____ of the body detect the stimulus. This event is signaled to the (b) _____ of the brain. One of the resulting responses is by the blood vessels of the skin, which (c) _____ . This response (d) _____ heat. For the loss of heat from the body, the skin blood vessels (e) _____ . Another response is the activation of (f) _____ glands.

Answers:
1. epithelial—covers body surfaces, lines cavities; connective—binds and supports body parts; muscular—movement of body parts; nervous—transmits impulses
2. a, b, c
3. a. beneath skin, in most organs; protective covering
   b. beneath skin, around kidneys, surface of heart; insulates and provides padding
   c. tendons, ligaments; connects muscles to bones, connects bones to bones
   d. ends of long bones, external ear and nose, intervertebral disks; protection and support
   e. bones of the body; compact bone—mechanical strength, spongy bone—makes red blood cells
   f. blood of the body; red blood cells carry oxygen, white blood cells fight infection, platelets are involved in blood clotting
4. a. striated b. striated c. smooth d. striated e. smooth
5. a. dendrites to cell body to axon b. send signals c. glial d. protection, support, and nutrition
6. a. circulatory b. digestive c. respiratory d. skeletal e. endocrine
7. a. dorsal b. ventral c. dorsal d. ventral e. ventral
8. It maintains the characteristics of the body's internal environment within a normal range.
9. a, b, c
10. a. receptors b. hypothalamus c. constrict d. conserves e. dilate f. sweat

CRITICAL THINKING QUESTIONS

1. How is each level of organization of the human body more than merely the sum of its parts?
2. Why is it important to maintain a relatively constant environment?

Answers:
1. The parts are integrated, interacting to form a more complex structure. For example, a tissue is more than just a group of cells.
2. The constant environment in each case refers to an optimal condition at which the body functions best, such as temperature or glucose concentration in the blood.

CHAPTER TEST

Indicate whether each of the following statements is
a. true.
b. false.

_____ 1. Columnar cells have a flat shape.
_____ 2. Endocrine glands secrete products into ducts.
_____ 3. Squamous cells have a long, pillar shape.
_____ 4. Much more of the makeup of connective tissue is matrix rather than cells.
_____ 5. Fibroblasts store fat in adipose tissue.
_____ 6. Loose connective tissue lies beneath the epithelium of the skin.
_____ 7. Tendons connect bone to bone.
_____ 8. The ends of long bones contains spongy bone.
_____ 9. The Haversian canal is part of the microscopic makeup of compact bone.
_____ 10. The plasma is the cellular portion of the blood.
_____ 11. The axial skeleton contains the bones of the limbs.
_____ 12. Cardiac and smooth muscle make up the internal organs of the body.
_____ 13. The axon sends a signal toward the cell body of the same neuron.
_____ 14. The epidermis is the deeper, thicker layer of the skin.
_____ 15. Positive feedback involves a response that tends to cancel the stimulus that brought on the response.

Answers: 1. b 2. b 3. b 4. a 5. a 6. a 7. b 8. a 9. a 10. b 11. b 12. a 13. b 14. b 15. b

CHAPTER

16

Circulation, Blood, and Immunity

CHAPTER CONCEPTS

1. The purpose of circulation is to deliver blood to the capillaries, where exchange of molecules takes place.
2. All of the functions of blood can be correlated with the ability of the body to maintain a relatively constant internal environment.
3. Immunity protects the body from disease and includes general and specific mechanisms.

CHAPTER REVIEW

Among invertebrates, cnidarians and flatworms supply cells through diffusion from a gastrovascular cavity. An internal transport system is lacking. Other invertebrates have a circulatory system of transport, which may be either internal or external.

Among vertebrates, fish have a single circulatory loop working with the gills. The other vertebrates have pulmonary and systemic circulation. In either type of circuit, blood flows from the pumping action of the heart through the following series of vessels: arteries, arterioles, capillaries, venules, and veins. Capillaries are the exchange vessels. Blood is forced along the vessels by alternating systolic and diastolic pressure. The oxygenation of blood is most efficient in the 4-chambered heart of birds and mammals.

In humans, the heartbeat originates with the SA node of the conduction system, which stimulates the atria. The AV node signals the ventricles. The action of valves prevents the backward flow of blood through this pump. The cardiac cycle is marked by the systole and diastole of the atria and ventricles.

Unlike the circulatory system, the lymphatic system is a one-way system. It contains lymphatic vessels and capillaries only, which return a derivative of the tissue fluid (lymph) to the systemic blood flow. T cells of this system provide cellular immunity.

The blood consists of the plasma and cells. The plasma is mostly water with dissolved substances. Red blood cells transport oxygen. Several types of white blood cells fight infection and establish immunity. One type, neutrophils, have phagocytic ability. B lymphocytes carry out antibody-mediated immunity. Platelets promote the clotting of blood. Their activity leads to the formation of fibrin, which traps red blood cells, forming the clot.

Blood can be typed by the ABO system. There are 4 types, based on the antigen content of the blood: A, B, AB, and O.

VOCABULARY

The list in the first column includes some of the chapter's terms. Each term is followed by the page where it first appears. Locate each term in the chapter and read its description. Then match the meanings to the terms.

1. blood (p. 245)
2. capillary (p. 245)
3. diastole (p. 244)
4. hemoglobin (p. 249)
5. neutrophil (p. 250)
6. platelet (p. 250)
7. pulmonary (p. 242)
8. SA node (p. 244)
9. systole (p. 244)

a. contraction of the heart
b. circulation with the lungs
c. starts blood clotting
d. pigment
e. exchange vessel
f. pacemaker
g. cells plus plasma
h. type of white blood cell
i. relaxation of the heart

Answers: 1. g 2. e 3. i 4. d 5. h 6. c 7. b 8. f 9. a

LEARNING ACTIVITIES

Study the text section by section as you answer the following questions.

Open and Closed Circulatory Systems (p. 242)

1. Label each of the following as describing a closed or an open circulatory system.
    a. found in mollusks _____
    b. found in annelids _____
    c. blood pumped by a heart _____
    d. blood not enclosed within vessels _____
    e. blood ebbs and flows within cavities _____
    f. blood never runs freely _____
2. Label each of the following statements as describing arteries, veins, or capillaries.
    a. carry blood away from the heart _____
    b. carry blood toward the heart _____
    c. deliver blood to the venules _____
    d. microscopic vessels _____
    e. have large number of valves _____
    f. have the thickest walls _____
    g. receive blood directly from the arterioles _____

Human Circulation (p. 243)

3. Label the following diagram of the heart.

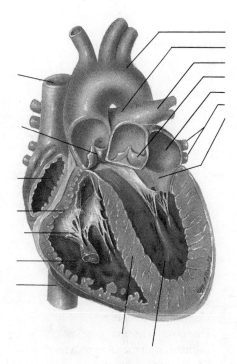

Copyright © Mark Lefkowitz

4. Use the numbers 1–10 to indicate the sequence of blood flowing through the circulatory system. Use the left ventricle as 1, the starting point.

   _____ a. left atrium
   _____ b. left ventricle
   _____ c. pulmonary artery
   _____ d. pulmonary vein
   _____ e. lungs
   _____ f. right atrium
   _____ g. right ventricle
   _____ h. systemic artery
   _____ i. systemic capillary
   _____ j. systemic vein

5. How does hepatic portal circulation differ from the systemic and pulmonary circuits?
6. The signal from the (a) _____ node initiates the contraction of the (b) _____ (chambers). This stimulus is picked up by the (c) _____ node, and this initiates the contraction of the (d) _____ (chambers). When the chambers are not contracting, they are relaxing. Contraction is termed (e) _____ , and resting is termed (f) _____ . Contraction of the atria forces the blood through the (g) _____ valves into the (h) _____ . The closing of these valves is the lub sound. Next, the ventricles contract and force the blood into the arteries. Now the (i) _____ valves close; this is the Dupp sound.

The Lymphatic System (p. 248)

7. Select the correct statements about the lymphatic system.
   a. The heart pumps lymph into the lymphatic capillaries.
   b. Lymphatic vessels have a construction similar to systemic veins.
   c. Lymph is eventually combined with the systemic venous blood.
   d. Lacteals absorb amino acids.
   e. The spleen is a lymphoid organ.

Human Blood (p. 249)

8. Label each of the following phrases with one of the following components of blood.

   red blood cells
   white blood cells
   platelets
   plasma

   a. _____ Activity leads to fibrin formation.
   b. _____ Basophil is an example.
   c. _____ Carry oxygen.
   d. _____ Contain hemoglobin.
   e. _____ Dissolves salts and proteins.
   f. _____ Formed from megakaryocytes.
   g. _____ Liquid portion.
   h. _____ Live about 120 days.
   i. _____ Monocyte is an example.
   j. _____ T cells and B cells.

9. List 3 steps leading to the clotting of the blood, in the order that they occur.

10. Label each of the following statements about capillary exchange as true or false.
    a. Water leaves the venous end of the capillary.
    b. Water enters the arterial end of the capillary.
    c. Blood pressure forces fluid out of the capillary.
    d. Osmotic pressure draws fluid back into the capillary.

11. Complete the following statements about the ABO blood types.
    a. Blood type A has antibody(s) _____ .
    b. Blood type B has antibody(s) _____ .
    c. Blood type AB has antigen(s) _____ .
    d. Blood type O has antibody(s) _____ .

Answers:
1. a. open b. closed c. closed d. open e. open f. closed
2. a. arteries b. veins c. capillaries d. capillaries e. veins f. arteries g. capillaries
3. see text fig. 16.3, p. 244
4. a. 10 b. 1 c. 7 d. 9 e. 8 f. 5 g. 6 h. 2 i. 3 j. 4
5. For the hepatic portal circuit, blood in the capillaries of several organs flows to a second capillary bed in the liver before its release into the systemic veins. The pulmonary circulation is between the lungs and heart. The systemic circulation serves most of the remainder of the body.
6. a. SA b. atria c. AV d. ventricles e. systole f. diastole g. AV h. ventricles i. semilunar
7. b, c, e
8. a. platelets b. white c. red d. red e. plasma f. platelets g. plasma h. red i. white j. white
9. platelets and tissue cells produce prothrombin activator; prothrombin is converted to thrombin; fibrinogen is converted to fibrin
10. a. false b. false c. true d. true
11. a. B b. A c. A, B d. A, B

CRITICAL THINKING QUESTIONS

1. What is the advantage of a closed circulatory system over an open circulatory system?
2. What do you think would happen to the heartbeat if the AV node were not stimulated by the SA node?

Answers:
1. The blood in the closed system is sent efficiently and directly to supply specific body regions.
2. The ventricles would not be signaled to contract and ventricular systole would not take place.

CHAPTER TEST

Indicate your answers by circling the letter. Do not refer to the text when taking this test.

1. Which organism transports by a gastrovascular cavity?
   a. cnidarian
   b. earthworm
   c. insect
   d. mollusk
2. Indicate the correct pathway of blood flow.
   a. arteries, capillaries, veins
   b. arteries, veins, capillaries
   c. veins, arteries, capillaries
   d. veins, capillaries, arteries
3. The function of the heart valves is to
   a. prevent the backward flow of blood.
   b. pump blood.
   c. separate the 2 sides of the heart.
   d. signal the chambers of the heart.
4. When the atria contract, the ventricles are in
   a. systole.
   b. diastole.
5. The lower pressure is _____ pressure.
   a. systolic
   b. diastolic
6. Select the *incorrect* phrase about red blood cells.
   a. contain hemoglobin
   b. contain iron
   c. respond during inflammation
   d. transport oxygen
7. Select the *incorrect* phrase about white blood cells.
   a. activate prothrombin
   b. exist in agranular and granular forms
   c. lymphocyte is one type
   d. neutrophil is one type
8. For the coagulation of blood, fibrinogen is converted to
   a. calcium.
   b. fibrin.
   c. prothrombin.
   d. thrombin.
9. Type B blood contains _____ antigen(s) and _____ antibody(s).
   a. 0, 1
   b. 1, 0
   c. 1, 1
   d. 2, 2
10. Lymph is a derivative of the
    a. blood cells.
    b. tissue fluid.

Answers: 1. a 2. a 3. a 4. b 5. b 6. c 7. a 8. b 9. c 10. b

CHAPTER 17

Digestion, Respiration, and Excretion

CHAPTER CONCEPTS

1. The digestive, respiratory, and excretory systems help maintain homeostasis.
2. The digestive system supplies an animal with nutrient molecules.
3. The respiratory system carries on gas exchange.
4. The excretory system rids the body of metabolic wastes and excess water; it also maintains the pH.

CHAPTER REVIEW

As feeders, animals can be continuous or discontinuous. A complete digestive tract has specialized compartments.

The teeth are one important digestive structure in mammals. The teeth have several different shapes, each specialized for a different function.

Human digestion begins with chewing in the mouth. Here salivary amylase initiates the chemical digestion of carbohydrates. The chemical digestion of proteins begins in the stomach through the action of the enzyme pepsin. Food is also stored there. The chemical and physical breakdown of food produces chyme. The action of other enzymes and bile in the small intestine finishes the chemical breakdown of all food molecules. The products of chemical digestion are absorbed in this region. Water and minerals are reabsorbed in the large intestine. Wastes are eliminated from the body here.

Several structures near the small intestine produce substances that are secreted into the digestive tract and influence its activity: the liver makes bile, the gallbladder stores bile, and the pancreas produces enzymes.

Along with ingested carbohydrates, proteins, and fats, minerals and vitamins contribute to the well-balanced diet of an individual.

For some aquatic organisms, the entire body surface functions for gas exchange. More advanced animals have evolved specialized respiratory organs. Gills (aquatic) and tracheal systems (terrestrial) are examples. Lungs are another gas exchange structure among some species of land animals and birds have evolved a series of air sacs for the one-way flow of air.

The breathing cycle of humans begins by the contraction of the diaphragm and rib-elevating muscles. Their contraction creates a negative pressure, drawing air into the lungs. A reverse of this process drives air out of the lungs. As air is inhaled, it passes through a sequence of structures of the respiratory tract: nasal cavity, pharynx, larynx, trachea, bronchi, bronchioles, and alveoli.

Once in the alveoli of the lungs, carbon dioxide and oxygen are exchanged with the blood by diffusion. Diffusion also serves the cells throughout the body with oxygen while also being responsible for the liberation of carbon dioxide from body cells. Oxygen is mainly transported in the blood through association with hemoglobin. Carbon dioxide is mainly transported as part of the bicarbonate ion.

The form by which animals excrete nitrogen depends on their type of environment: usually ammonia in water and either urea or uric acid on land. Both freshwater and saltwater fishes face problems of salt and water balance. Several intake and output mechanisms maintain optimal body concentrations of these substances.

Different groups of animals have evolved different excretory organs. Humans have evolved a kidney, consisting of microscopic units called nephrons. Through the action of the nephrons, the kidneys control the chemistry of the blood through pressure filtration, selective reabsorption, and tubular secretion. Through these processes, humans excrete a hypertonic urine. ADH regulates water reabsorption and another hormone, aldosterone, controls the reabsorption of sodium.

The kidneys also control the pH of the blood by regulating the hydrogen ion concentration.

The artificial kidney machine, via dialysis, replaces normal kidney function to control the concentration of substances in the blood.

VOCABULARY

The list in the first column includes some of the chapter's terms. Each term is followed by the page where it first appears. Locate each term in the chapter and read its description. Then match the meanings to the terms.

1. aldosterone (p. 272)
2. amylase (p. 263)
3. colon (p. 259)
4. glomerulus (p. 270)
5. larynx (p. 266)
6. liver (p. 262)
7. medulla (p. 270)
8. peristalsis (p. 259)
9. trachea (p. 266)
10. trypsin (p. 263)

a. enzyme working on proteins
b. large intestine
c. inner layer of kidney
d. voice box
e. enzyme working on starch
f. makes bile
g. part of nephron
h. hormone
i. windpipe
j. rhythmic contraction

Answers: 1. h 2. e 3. b 4. g 5. d 6. f 7. c 8. j 9. i 10. a

LEARNING ACTIVITIES

Study the text section by section as you answer the following questions.

Procuring and Digesting Food (p. 258)

1. Select the correct statements.
    a. Some terrestrial animals are filter feeders.
    b. Most animals are discontinuous feeders.
    c. The planarian has a complete digestive tract.
    d. The earthworm has an incomplete digestive tract.
    e. Molars are specialized for grinding and crushing food.
2. Write 1–6 in the following blanks to indicate the order in which food passes through the digestive tract.

    _____ esophagus
    _____ large intestine
    _____ mouth
    _____ pharynx
    _____ small intestine
    _____ stomach

3. Refer to the drawing of the human digestive system and indicate where each of the following occurs by using letters from the figure.
   a. Food here is prevented from entering the trachea.
   b. Absorption of nutrients occurs here.
   c. Chyme passes into this region.
   d. HCl is secreted here.
   e. Large quantities of food are stored here.
   f. The chemical breakdown of carbohydrates begins here.
   g. The chemical breakdown of lipids begins here.
   h. The chemical breakdown of proteins begins here.
   i. The salivary glands are found here.
   j. The reabsorption of water and minerals occurs here.
   k. This structure conducts food to the stomach.
   l. This structure makes bile.
   m. This structure secretes enzymes into the small intestine.
   n. This structure stores bile.
   o. Villi are found on the inside surface of this structure.

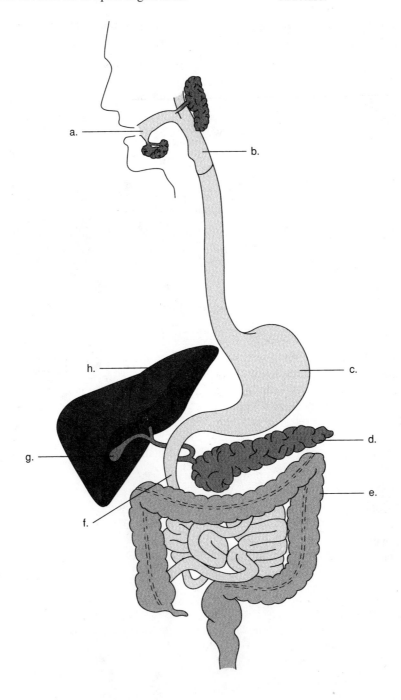

4. For each of the following enzymes, name the region where it functions in the human digestive tract and include the chemical reaction that it produces.
   a. lipase _____
   b. pancreatic amylase _____
   c. pepsin _____
   d. salivary amylase _____
   e. trypsin _____
5. List the final molecules produced by enzymatic hydrolysis of each of the following products of digestion.
   a. peptides _____
   b. maltose _____
6. Select the correct statements.
   a. The new pyramid from the Agriculture Department is built on a base of grains.
   b. Dietary fat has been implicated in cancer of the colon.
   c. Simple carbohydrates are empty calories.
   d. A 154-lb man requires 80 g of protein per day.
   e. The insoluble fiber of wheat bran has a high nutritional value.
   f. Dairy products are at the top of the new pyramid.

Carrying on Gas Exchange (p. 264)

7. Select the correct statements.
   a. Small organisms have specialized organs for gas exchange.
   b. The planarian uses its body surface for gas exchange.
   c. Tracheae are aquatic respiratory organs.
   d. Birds do not retain air in their lungs after expiration.
   e. The internal lining of the lungs is most divided in the mammals.
   f. The elevation of the diaphragm drives air out of the lungs.
8. Rearrange the following structures to indicate the pathway taken by inhaled air: alveolus, bronchiole, bronchus, larynx, nose, pharynx, trachea.

9. From #8, name the structure that functions for each of the following:
   a. gas exchange with blood flowing to and from the heart _____
   b. sound production _____
   c. filtering, moistening, and warming the air _____
   d. conduction of air into the thoracic cavity _____
   e. passage of air into the lungs _____
10. Indicate whether each of the following phrases is an example of inhalation or exhalation.
    a. lungs expanded _____
    b. diaphragm and rib muscles elevated _____
    c. diaphragm flattens _____
    d. enlarged chest _____
    e. air pressure in lungs is less than that in the outside environment
11. a. What basic transport process moves oxygen and carbon dioxide at the alveolus?
    b. How is oxygen mainly transported by the blood?
    c. How is carbon dioxide mainly transported by the blood?

Excreting Metabolic Wastes (p. 269)

12. Select the correct statements.
    a. Ammonia is a main nitrogen waste product for aquatic organisms.
    b. Uric acid is the least toxic nitrogen waste product.
    c. Freshwaster fish face a problem of water gain.
    d. Marine fish face a problem of water loss.

13. Indicate the four organs of the urinary system in the order that urine passes through them when excreted.

    _____ (first)

    _____ (last)

14. Write 1–6 to indicate the direction followed by an excreted molecule through the nephron.

    _____ Bowman's capsule

    _____ collecting duct

    _____ distal convoluted tubule

    _____ glomerulus

    _____ loop of Henle

    _____ proximal convoluted tubule

15. Provide definitions for
    a. pressure filtration.
    b. tubular reabsorption.

16. How does the kidney regulate the concentration of each of the following in the blood?
    a. water
    b. sodium
    c. hydrogen ions

Answers:

1. b, e
2. esophagus—3, large intestine—6, mouth—1, pharynx—2, small intestine—5, stomach—4
3. a. b b. f c. f d. c e. c f. a g. f h. c i. a j. e k. b l. h m. d n. g o. f
4. see text table 17.2, p. 263
5. a. amino acids b. glucose
6. a, b, c, f
7. b, d, e, f
8. nose (first), pharynx, larynx, trachea, bronchus, bronchiole, alveolus (last)
9. a. alveoli b. larynx c. nose d. trachea e. bronchus
10. a. inhalation b. exhalation c. inhalation d. inhalation e. inhalation
11. a. diffusion b. combined with hemoglobin c. as part of the bicarbonate ion
12. a, b, c, d
13. kidney (first), ureter, bladder, urethra (last)
14. Bowman's capsule—2, collecting duct—6, distal convoluted tubule—5, glomerulus—1, loop of Henle—4, proximal convoluted tubule—3
15. a. the movement of substances from the glomerulus into the Bowman's capsule
    b. the return of substances from the tubule into the blood of the peritubular capillaries
16. a. After filtration, water is reabsorbed mainly in the proximal convoluted tubule as needed. ADH controls the final stages of reabsorption.
    b. Sodium is reabsorbed after being filtered. Aldosterone affects the rate of reabsorption.
    c. The kidney secretes H^+ near the end of the tubule.

CRITICAL THINKING QUESTIONS

1. The well-preserved skull of a mammal is found, showing well-developed molars and poorly developed canine teeth. What can you conclude about the probable diet of this organism?
2. How do the alveoli exemplify structures that are well adapted to their function?

Answers:

1. Molars are adapted for grinding, the kind of action needed to break down plant food. Canine teeth are needed more for the tearing of flesh.
2. They are numerous, providing a greater total surface area for gas exchange. Their moist, thin surfaces also maximize the diffusion for gas exchange.

CHAPTER TEST

Indicate your answers by circling the letter. Do not refer to the text when taking this test.

1. Salivary amylase speeds up the conversion of
   a. protein to amino acids.
   b. protein to peptides.
   c. starch to glucose.
   d. starch to maltose.
2. Select the correct sequence for the passage of food.
   a. pharynx, esophagus, stomach
   b. large intestine, small intestine, stomach
   c. pharynx, stomach, esophagus
   d. stomach, large intestine, small intestine
3. Villi serve to
   a. increase surface area for absorption.
   b. increase the synthesis of enzymes.
   c. speed up elimination of wastes.
   d. speed up loss of water from the body.
4. Pepsin works best in an environment that is
   a. acidic.
   b. basic.
5. Dietary _____ has been implicated in cancer of the colon.
   a. starch
   b. glucose
   c. fat
   d. protein
6. A major adaptation for gas exchange in planarians is the
   a. countercurrent exchange in the lungs.
   b. lack of a gastrovascular cavity.
   c. presence of a thin, flattened body.
   d. thin surface with parapodia.
7. The inner lining of the lungs is most divided in the
   a. birds.
   b. amphibians.
   c. reptiles.
   d. mammals.
8. Inhalation in the human body occurs by the _____ of the diaphragm and the _____ of the ribs.
   a. elevation, dropping
   b. elevation, elevation
   c. flattening, dropping
   d. flattening, elevation
9. Air is exhaled by passing through this order of structures in the respiratory tract.
   a. alveolus, bronchiole, bronchus, pharynx
   b. bronchus, bronchiole, trachea, pharynx
   c. pharynx, larynx, trachea, bronchus
   d. trachea, alveolus, bronchus, bronchiole
10. Most carbon dioxide in the blood is transported in the
    a. bicarbonate ion.
    b. carbon dioxide molecule.
    c. hemoglobin molecule.
    d. oxygen molecule.
11. Marine fish face the problem of water loss from the body.
    a. true
    b. false
12. The innermost hollow chamber of the kidney is the
    a. cortex.
    b. glomerulus.
    c. medulla.
    d. pelvis.
13. Pressure filtration occurs in the _____ ; reabsorption occurs in the _____ .
    a. Bowman's capsule, nephron tubule
    b. nephron tubule, Bowman's capsule
14. By selective reabsorption, substances pass from the
    a. blood into the tubule.
    b. tubule into the blood.
15. The hormone aldosterone promotes the
    a. excretion of potassium only.
    b. reabsorption of sodium only.
    c. excretion of potassium and reabsorption of sodium.
    d. elimination of water.

Answers: 1. d 2. a 3. a 4. a 5. c 6. c 7. d 8. d 9. a 10. a 11. a 12. d 13. a 14. b 15. c

CHAPTER 18

Nervous and Endocrine Systems

CHAPTER CONCEPTS

1. The nervous and endocrine systems coordinate the activity of all systems of the body.
2. The evolutionary history of organisms shows that, over time, there was an increase in the complexity of the nervous system.
3. In complex animals, the ability to respond is dependent upon sense organs, the nervous system, and effectors, including muscle effectors.
4. Sense organs respond to only certain types of stimuli.
5. Movement requires a source of energy; muscles use ATP energy.

CHAPTER REVIEW

Working along with the endocrine system, the nervous system regulates and coordinates a variety of body functions. The further up the phylogenetic tree of the animal kingdom, the greater the comparative complexity of the nervous system.

The neuron is the impulse-sending cell of the nervous system. It sends an electrical signal through the dendrites, cell body, and axon. This signal involves the shifting of sodium and potassium ions across the nerve cell membrane. By these ionic shifts, a resting membrane potential is changed to an action potential, the nerve impulse. The nerve fibers that fire most rapidly are myelinated with nodes.

Transmission across the synapse, by contrast, is chemical. It depends on the release and diffusion of a neurotransmitter from a presynaptic cell to excite a postsynaptic cell.

The peripheral nervous system has 2 branches—the somatic and autonomic. The autonomic nervous system controls motor function of internal organs by its sympathetic and parasympathetic branches.

Neurons are classified as sensory neurons, motor neurons, or interneurons, all of which are activated in a reflex arc.

Anatomically, the central nervous system consists of the brain and spinal cord. Humans have the most well-developed cerebral cortex. The lobes of the cortex can be mapped for sensory and motor function.

Endocrine glands in humans secrete hormones into the bloodstream. By the circulation of blood, they reach target organs where they produce a response.

Three categories of chemical messengers exist in biological systems: (1) pheromones, working between organisms, (2) endocrine hormones, working within an organism and (3) local messengers, working between adjacent cells in the organism.

Receptors are specialized cells that detect environmental changes or stimuli. Several types of sense organs are present in humans, including mechanoreceptors (hearing) and photoreceptors (vision). Effectors, such as the skeletal muscles, respond to environmental stimuli.

VOCABULARY

The list in the first column includes some of the chapter's terms. Each term is followed by the page where it first appears. Locate each term in the chapter and read its description. Then match the meanings to the terms.

1. ACh (p. 280)
2. autonomic (p. 283)
3. hormone (p. 286)
4. myosin (p. 291)
5. neuron (p. 277)
6. pheromone (p. 286)
7. sclera (p. 288)
8. stapes (p. 290)
9. temporal (p. 285)
10. thalamus (p. 285)

a. neurotransmitter
b. middle ear structure
c. lower brain region
d. lobe of cerebral cortex
e. environmental hormone
f. signal in the blood
g. outer layer of the eye
h. muscle protein
i. branch of PNS
j. cell sending message

Answers: 1. i 2. f 3. h 4. j 5. a 6. e 7. g 8. c 9. d 10. b

LEARNING ACTIVITIES

Study the text section by section as you answer the following questions.

The Nervous System (p. 277)

1. Select the correct statements.
   a. The nervous system of the hydra resembles a network of threads.
   b. The planarian has the rudiments of a CNS and PNS.
   c. The nervous system of an annelid is typical of an invertebrate.
   d. An insect's nervous system contains about one million neurons.
   e. In fish and amphibians, the forebrain has largely an auditory function.
2. Complete the following statements.
   a. The _____ of a neuron sends information away from the cell body.
   b. The _____ of a neuron sends information toward the cell body.
   c. The cell body of a neuron contains the _____ and organelles.
3. Select the correct statements about the nerve impulse and transmission at the synapse.
   a. At rest, the inside of the neuron is more negative than the outside.
   b. Myelination and nodes on the neuron slow the rate of conduction.
   c. Repolarization of the neuron requires the opening of potassium gates.
   d. The action potential depends on an influx of sodium ions into the cell.
   e. There is normally an equal distribution of potassium ions on both sides of the nerve cell membrane.
   f. There is normally an equal distribution of sodium ions on both sides of the nerve cell membrane.
   g. The nature of the synaptic signal is chemical.
   h. At the synapse, a postsynaptic cell signals a presynaptic cell.
4. Use letters a–e to indicate the chronological order in which a signal passes through the reflex arc.

   _____ interneuron
   _____ motor neuron
   _____ receptor
   _____ sensory neuron
   _____ skeletal muscle

5. Complete the following statements.
   a. The central nervous system consists of the _____ and _____ .
   b. The 2 branches of the peripheral nervous system are the _____ and _____ .
   c. The 2 branches of the autonomic nervous system are the _____ and _____ .

85

6. Fill in the function of the parts of the brain listed.

   Brain Part **Function**
   a. cerebrum
   b. thalamus
   c. hypothalamus
   d. cerebellum
   e. medulla oblongata

The Endocrine System (p. 286)

7. Select the correct statements about the endocrine system.
   a. A pheromone is transported in the blood.
   b. The endocrine system acts more slowly than does the nervous system.
   c. The hypothalamus signals the pituitary gland.
   d. Hormones are signals between cells.
   e. Calcitonin raises the level of calcium in the blood.
   f. The adrenal cortex causes gluconeogenesis.
   g. Thyroxin increases the metabolic rate.
   h. The anterior pituitary secretes ADH.

Receptors and Effectors (p. 287)

8. State the structure and function for each part of the eye indicated in the figure.

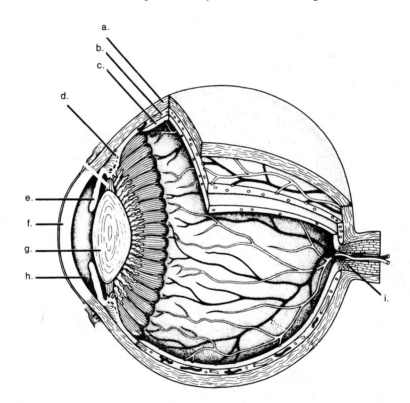

9. Select the correct statements about the physiology of vision.
   a. The cornea is the first structure of the eye to refract light rays.
   b. With normal aging, the ability of the lens to accommodate close objects improves.
   c. The rods of the retina are active in bright light.
   d. The cones of the retina are active in dim light.
   e. Color vision depends on the activity of the cones.
   f. The rods contain rhodopsin.

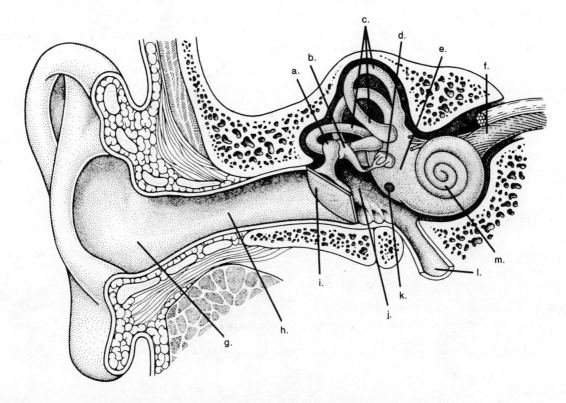

10. On the lines below, state the structure and function for each part of the ear.
11. Write 1–7 to indicate the direction taken by the signal for hearing.

   _____ auditory canal
   _____ cochlea
   _____ incus
   _____ malleus
   _____ pinna
   _____ stapes
   _____ tympanic membrane

12. State the function of these inner ear structures.
    a. semicircular canals
    b. vestibule
13. Select the correct statements about muscle effectors.
    a. The fibers are the cells of the skeletal muscle.
    b. Actin and myosin are proteins responsible for muscle contraction.
    c. The sarcomere is another name for the entire muscle.
    d. By the sliding filament theory, myosin filaments slide past the actin.
    e. ATP supplies the energy for muscle contraction.
    f. Calcium is stored in the sarcoplasmic reticulum.

Answers:
1. a, b, c, d
2. a. axon b. dendrites c. nucleus
3. a, c, d, g
4. 3—interneuron, 4—motor neuron, 1—receptor, 2—sensory neuron, 5—skeletal muscle
5. a. brain, spinal cord, b. somatic, autonomic c. sympathetic, parasympathetic
6. a. consciousness and all higher activities like memory and language; reception of sensory data and initiation of motor responses
   b. gatekeeper of the brain; alerts higher centers to receive information
   c. brings about homeostasis; has centers for hunger, sleep, and body temperature
   d. motor coordination
   e. control of internal organs

7. b, c, d, f, g
8. a. sclera—protection b. choroid—absorbs light rays c. retina—contains sense receptors for sight d. ciliary muscle—adjusts the shape of the lens e. iris—regulates the entrance of light f. cornea—refraction of light g. lens—focusing h. anterior chamber containing aqueous humor—refraction of light rays i. optic nerve—transmission of nerve impulse to the brain
9. a, e, f
10. a. hammer—transmits vibrations b. anvil—transmits vibrations c. semicircular canals—dynamic equilibrium d. stirrup—transmits vibrations e. vestibule—static equilibrium f. auditory nerve—transmission of nerve impulse g. pinna—reception of sound waves h. auditory canal—collection of sound waves i. eardrum—starts vibration of ossicles j. middle ear—location ossicles k. round window—bulges to relieve fluid pressure on the inner ear l. Eustachian tube—connects middle ear to pharynx m. cochlea—contains hearing receptors
11. 2—auditory canal, 7—cochlea, 5—incus, 4—malleus, 1—pinna, 6—stapes, 3—tympanic membrane
12. a. dynamic equilibrium b. static equilibrium
13. a, b, e, f

CRITICAL THINKING QUESTIONS

1. Damage to any of 3 major components prevents vision in a person. Only one is a sensory organ. What are all 3 and why are they necessary?
2. How is the structure of the neuron well suited for its function?

Answers:
1. The receptor cells of the retina detect stimuli. A sensory pathway delivers a signal to the brain. The brain must interpret this input in order for the person to see.
2. It is a long, thin cell that reaches over distances in the body and is capable of rapidly delivering signals to a body region.

CHAPTER TEST

Indicate your answers by circling the letter. Do not refer to the text when taking this test.

1. Within a neuron, a signal travels from the
    a. cell body to dendrites to axon.
    b. cell body to axon to dendrites.
    c. dendrites to axon to cell body.
    d. dendrites to cell body to axon.
2. Relative to the inside, the outside of a neuron at rest is
    a. negative.
    b. positive.
3. The production of an action potential begins with the entry of _____ into the cell.
    a. potassium
    b. sodium
4. Which is the last step for transmission at the synapse?
    a. neurotransmitter diffuses across the synapse
    b. neurotransmitter is released from vesicles
    c. postsynaptic cell is excited
    d. synaptic vesicles rupture
5. A reflex arc begins with a(n) _____ and ends with a(n) _____ .
    a. effector, receptor
    b. receptor, effector
6. Which is *not* a lobe of the cerebral cortex?
    a. frontal
    b. portal
    c. temporal
    d. occipital

7. Which region of the brain contains the center for the heartbeat?
    a. hypothalamus
    b. thalamus
    c. medulla oblongata
    d. cerebral cortex
8. Which hormone lowers the level of sugar in the blood?
    a. ADH
    b. insulin
    c. thyroxin
    d. ATP
9. Which endocrine gland produces a hormone that regulates calcium?
    a. pancreas
    b. parathyroid
    c. pituitary
    d. adrenal
10. Select the *incorrect* association.
    a. choroid—absorbs light rays
    b. cornea—refracts light rays
    c. pupil—admits light
    d. sclera—transmits impulses
11. The _____ of the retina detect colors; the _____ of the retina function in dim light.
    a. cones, rods
    b. rods, cones
12. Light arrives at the retina by passing through this sequence of structures.
    a. aqueous humor, cornea, lens, vitreous humor
    b. cornea, aqueous humor, lens, vitreous humor
    c. lens, cornea, aqueous humor, vitreous humor
    d. vitreous humor, lens, cornea, aqueous humor
13. Select the largest structure.
    a. actin
    b. muscle fiber
    c. myofibril
    d. sarcomere
14. By the sliding filament theory
    a. actin cross-bridges slide myosin.
    b. myosin cross-bridges slide actin.
15. The role of creatine phosphate is to
    a. slide filaments.
    b. store energy.
    c. bind cross-bridges.
    d. act as a protein.

Answers: 1. d 2. b 3. b 4. c 5. b 6. b 7. c 8. b 9. b 10. d 11. a 12. a 13. b 14. b 15. b

CHAPTER 19

Animal Reproduction and Development

CHAPTER CONCEPTS

1. Reproductive strategies are adapted to the environment.
2. Reproduction is under hormonal, rather than nervous, control.
3. Embryonic development provides evidence that all animals are related.
4. Embryonic development is an orderly series of stages.
5. Development occurs throughout the life of an animal.

CHAPTER REVIEW

Asexual reproduction involves one parent producing offspring with the same genotype and phenotype. Sexual reproduction produces genetic variations from the combination of gametes from 2 parents.

Sex cell production in the human male occurs in the seminiferous tubules of the testis. This process is regulated by FSH. Interstitial cells in the testis produce testosterone. This production is regulated by LH. Sperm cells mature in the epididymis and pass through the vas deferens and urethra during ejaculation. The seminal vesicles, prostate gland, and Cowper's glands add seminal fluid to this pathway.

In females, egg production occurs in the ovary. Fertilization of a secondary oocyte, released from the ovary into the oviduct during ovulation, produces a fertilized egg. The follicle, remaining in the ovary and eventually converted to a corpus luteum, produces estrogen and progesterone through the ovarian and uterine cycles.

The oviduct leads to the uterus, which opens into the vagina. If fertilization occurs, the developing embryo implants onto the thickened lining of the uterus.

Development involves the differentiation and morphogenesis of cells. Development of the lancelet and human, for example, differs due to the amount of yolk in the egg.

The establishment of 3 germ layers in the gastrula leads to the formation of organs. Morphogenesis depends on environmental factors as well as genetic ones.

Human development consists of the embryonic and fetal stages. Early events include fertilization in the oviduct, cleavage of the early embryo in the oviduct, morula formation, blastocyst formation, gastrulation, and formation of 3 germ layers.

Several extraembryonic membranes function in human development, including the chorion and the amnion.

VOCABULARY

The list in the first column includes some of the chapter's terms. Each term is followed by the page where it first appears. Locate each term in the chapter and read its description. Then match the meanings to the terms.

1. budding (p. 298)
2. corpus luteum (p. 302)
3. differentiation (p. 305)
4. endometrium (p. 300)
5. epididymis (p. 299)
6. estrogen (p. 301)
7. morphogenesis (p. 305)
8. oviduct (p. 300)
9. prostate (p. 299)
10. testosterone (p. 300)

a. male sex hormone
b. female sex hormone
c. cell specialization
d. sperm cells mature here
e. cells establish a form
f. female tract structure
g. makes seminal fluid
h. lining of uterus
i. converted follicle
j. asexual reproduction

Answers: 1. j 2. i 3. c 4. h 5. d 6. b 7. e 8. f 9. g 10. a

LEARNING ACTIVITIES

Study the text section by section as you answer the following questions.

Introduction (p. 297)

1. Label each of the following as describing asexual reproduction or sexual reproduction.
   a. Budding is one type.
   b. Gametes are produced.
   c. Offspring have a different combination of genes than either parent.
   d. Offspring tend to have the same genotype and phenotype as the parents.
   e. Practiced by hermaphrodites.
   f. Regeneration is one type.

Human Reproduction (p. 297)

2. Using the blanks provided, identify and state a function for the parts of the human male urinary and reproductive system.

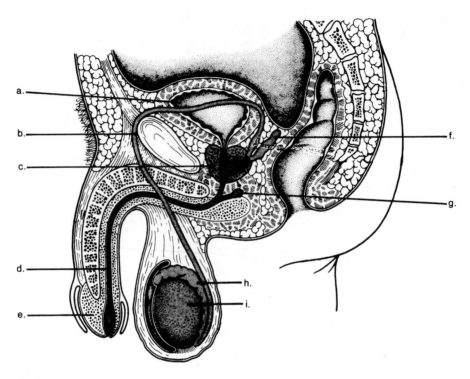

Name   **Function**

a.
b.
c.
d.
e.
f.
g.
h.
i.

3. Which of the structures in #2 add seminal fluid to the male reproductive tract?
4. Explain the function of the following male hormones.
   a. seminiferous tubules
   b. interstitial cells

5. Using the blanks provided in this diagram, identify and state a function for the parts of the human female reproductive and urinary structures.

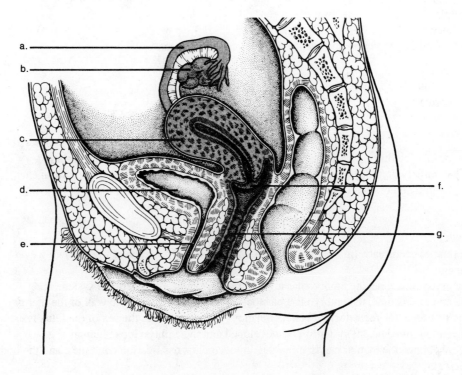

Name **Function**

a.
b.
c.
d.
e.
f.
g.

6. Select the correct statements about the ovarian cycle, uterine cycle, and female sex hormones.
   a. Only a small number of the follicles in the ovary ever mature.
   b. Ovulation is the union of the sperm cell and oocyte.
   c. During the ovarian cycle, the corpus luteum is converted into a follicle.
   d. There is a surge of LH at about the midpoint of the ovarian cycle.
   e. LH promotes the development of the corpus luteum.
   f. A rise in progesterone feeds back to diminish the levels of LH.
   g. The level of FSH is high during the first 5 days of the uterine cycle.
   h. There is an increase of estrogen during days 6–13 of the uterine cycle.
   i. Estrogen is largely responsible for secondary sex characteristics.
   j. Progesterone is partly responsible for breast development.

Development (p. 305)

7. Define the following.
   a. differentiation
   b. morphogenesis
8. Write 1–5 to indicate the correct sequence of developmental stages.

   _____ blastula
   _____ gastrula
   _____ morula
   _____ neurula
   _____ zygote

9. What determines the difference in the cleavage pattern of the lancelet and human embryo?
10. Name the 3 primary germ layers of the embryo, along with one structure developing from each.
11. Describe the function of each of the following extraembryonic membranes.
    a. allantois
    b. amnion
    c. placenta
12. Note the time of occurrence (in months or weeks when possible) for each of the following events of prenatal development.
    a. By the end of this time, the embryo has a working heart and central nervous system.
    b. By the end of this time, the embryo is a blastocyst and is implanted in the wall of the uterus.
    c. During this time, the reproductive system develops to the point that the sex of the fetus is externally apparent.
    d. By the end of this time, all systems have developed and the embryo looks human.
    e. During this time, the mother can first feel fetal movements, ossification continues, and the body is covered with fine hair and waxy secretions.

Answers:
1. a. asexual b. sexual c. sexual d. asexual e. sexual f. asexual
2. a. bladder, stores urine b. vas deferens, stores and conducts sperm c. prostate gland, secretes seminal fluid d. urethra, conducts urine and sperm e. penis, organ of copulation f. seminal vesicles, secrete seminal fluid g. Cowper's gland, secretes seminal fluid h. epididymis, storage and maturation of sperm i. testis, produces sperm and testosterone
3. seminal vesicles, prostate gland, Cowper's gland
4. a. carry out spermatogenesis b. produce testosterone
5. a. oviduct, conduction of the egg b. ovary, production of the egg and hormones c. uterus, organ where development occurs d. bladder, storage of urine e. urethra, conduction of urine f. cervix, opening of the uterus g. vagina, birth canal and organ of copulation
6. a, d, e, f, h, i, j
7. a. cells specialize for structure and function b. cells establish a shape and form
8. 3—blastula, 4—gastrula, 2—morula, 5—neurula, 1—zygote
9. the distribution of yolk
10. ectoderm—skin, nervous system; endoderm—lining of digestive tract, lungs, liver, pancreas; mesoderm—muscles, skeleton, kidney, circulatory system, gonads
11. a. collects nitrogenous wastes b. contains amniotic fluid to bathe embryo c. exchange barrier
12. see text table 19.5, p. 312

CRITICAL THINKING QUESTIONS

1. How is endocrine control of the female reproductive cycle more adaptive than nervous control?
2. How is sexual reproduction advantageous as a strategy for survival compared to asexual reproduction?

Answers:
1. Hormonal signaling is more deliberate, with longer-lasting effects.
2. Sexual reproduction promotes genetic variability, offering more alternatives for survival in a changing environment.

CHAPTER TEST

In 1–8, match each reproductive structure to its correct description.

1. anterior pituitary
2. epididymis
3. interstitial cells
4. ovary
5. oviduct
6. seminiferous tubules
7. uterus
8. vagina

a. organ of copulation
b. secretes FSH
c. produces estrogen
d. sperm cell production
e. testosterone production
f. site of development
g. maturation of sperm cells
h. site of fertilization

In 9–15, indicate whether each of the statements is
a. true.
b. false.

_____ 9. Differentiation is a process whereby cells move to establish a form.
_____ 10. The amount and distribution of the yolk in an egg influences the cleavage pattern.
_____ 11. The gastrula follows the blastula during embryo formation.
_____ 12. The inner layer of the embryo is the ectoderm.
_____ 13. The lining of the digestive tract develops from the endoderm.
_____ 14. The amniotic fluid is a source of nutrition in the developing embryo.
_____ 15. Morphogenesis is the specialization of cells.

Answers: 1. b 2. g 3. e 4. c 5. h 6. d 7. f 8. a 9. b 10. a 11. a 12. b 13. a 14. b 15. b

CHAPTER 20

Animal Behavior

CHAPTER CONCEPTS

1. All behavior has a genetic basis, but no behavior is entirely genetically determined.
2. Behavior increases the individual fitness of the individual.
3. Animals communicate with one another in various ways.
4. Individuals live in a society when the benefits outweigh the costs.

CHAPTER REVIEW

Behavior involves all activities that allow an animal to meet its needs. To some degree all behavior patterns are inherited. As a product of evolution, they are adaptive. There are 2 main categories of patterns: innate and learned. One example of learned behavior is imprinting. Fixed-action patterns are an example of an innate pattern.

There are other recognized modes of learning. With habituation, the original response ceases when the same stimulus is presented repeatedly. In the case of classical conditioning, associative learning is involved. Operant conditioning occurs with constant reinforcement. Insight learning involves problem solving and reasoning. Various types of feeding behavior and territoriality are examples of behavior patterns that enhance fitness.

A society is a group of individuals belonging to the same species, organized in a cooperative manner. There are advantages and disadvantages to the members living in such a group. Members can act aggressively or altruistically. Communication in a society can include several types: chemical, tactile, visual, and auditory.

According to the tenets of sociobiology, behavior patterns in a society can be adaptive. If adaptive, they tend to be preserved by natural selection.

VOCABULARY

The list in the first column includes some of the chapter's terms. Each term is followed by the page where it first appears. Locate each term in the chapter and read its description. Then match the meanings to the terms.

1. communication (p. 326)
2. FAP (p. 321)
3. imprinting (p. 322)
4. society (p. 326)
5. territory (p. 324)
6. waggle (p. 326)

a. transmission of a signal
b. innate behavior
c. learned behavior
d. type of dance
e. defended area
f. group of individuals

Answers: 1. a 2. b 3. c 4. f 5. e 6. d

LEARNING ACTIVITIES

Study the text section by section as you answer the following questions.

Mechanisms of Behavior (p. 320)

1. Identify each of the following as indicating innate behavior or learned behavior.
    a. The fixed-action pattern is one example. _____
    b. It requires a sign stimulus. _____
    c. Lorenz studied this type of behavior. _____
    d. By definition, this type of behavior results from experience. _____
    e. The pattern has a definite genetic basis. _____
    f. Imprinting is an example. _____
2. What evidence exists to show that behavior is under physiological control?
3. Name the behavior described by each of the following.
    a. An animal learns to respond to an irrelevant stimulus. _____
    b. An original response to a repeated stimulus ceases. _____
    c. The learning is trial and error. _____
    d. Problems are solved using previous experience. _____

Natural Selection and Behavior (p. 323)

4. Select the correct statements among the following.
    a. Behavior patterns are adaptive, enhancing fitness.
    b. If adaptive, the energy benefit of food should be less than the energy cost of getting it.
    c. Animals tend to select habitats that increase their reproductive success.
    d. A territory is an area defended by an animal.
    e. Territoriality diminishes reproductive success.
    f. The reproductive behavior of organisms should increase individual fitness.
    g. Members of a dominance hierarchy have differing access to resources.

Societies (p. 326)

5. Label each of the following types of communication as being chemical, visual, tactile, or auditory.
    a. Male birds undergo a color change. _____
    b. Female moths secrete products from abdominal glands. _____
    c. Baby gulls peck at a parent's beak. _____
    d. Female baboons show that they are in estrus. _____
    e. Honeybees demonstrate the waggle dance. _____
    f. Pheromones are discharged. _____
    g. Male crickets have calls. _____
6. Select the correct statements.
    a. Altruism can be adaptive behavior.
    b. Kin selection decreases an animal's inclusive fitness.
    c. Sociobiologists interpret human behavior as adaptive.
    d. As a branch of biology, sociobiology does not incorporate the tenets of evolutionary theory.

Answers:
1. a. innate b. innate c. learned d. learned e. innate f. learned
2. The importance of testosterone in the behavior of the male ringdove has been proven. FSH signals reproductive behavior in the female of the species.
3. a. classical conditioning b. habituation c. operant conditioning d. insight
4. a, c, d, f, g
5. a. visual b. chemical c. tactile d. visual e. tactile f. chemical g. auditory
6. a, c

CRITICAL THINKING QUESTIONS

1. What evidence shows that behavior is inherited?
2. Why do you think that altruistic behavior may not exist?

Answers:
1. Offspring have been shown to exhibit innate behaviors indicative of each parent. The genetic component is adaptive, promoting survival among the offspring.
2. Although it may appear altruistic, it is adaptive to promote the survival and reproductive fitness of the species exhibiting it.

CHAPTER TEST

Indicate whether each of the following statements is
a. true.
b. false.

_____ 1. Behavior patterns are somewhat inherited and evolve just as other animal traits do.
_____ 2. In innate behavior, the environmental component is probably more influential.
_____ 3. In learned behavior, the genetic component is probably more influential.
_____ 4. Fixed-action patterns occur in response to a sign stimulus.
_____ 5. Imprinting is learned later in life.
_____ 6. Pheromones represent a type of chemical communication.
_____ 7. Learning by imprinting occurs during a critical period.
_____ 8. Insight is one specialized form of conditioned learning.
_____ 9. Territoriality and dominance hierarchies minimize competition in a society.
_____ 10. Learning is a change in behavior as a result of experience.

Answers: 1. a 2. b 3. b 4. a 5. b 6. a 7. a 8. b 9. a 10. a

CHAPTER 21

The Biosphere and Ecosystems

CHAPTER CONCEPTS

1. All species live in a community of the biosphere.
2. Chemicals cycle and energy flows through an ecosystem.
3. Organisms in an ecosystem are dependent on one another and the physical environment.
4. Human activities affect the natural way in which chemicals cycle and energy flows in an ecosystem.

CHAPTER REVIEW

Within the biosphere each major terrestrial community is called a biome. Terrestrial biomes are produced by succession. These biomes include the deserts, tundra, grasslands, chaparral, and forests (tropical, temperate).

Various populations in a community interact to form an ecosystem. Producers, consumers (herbivores, carnivores, and omnivores), and decomposers are part of the biotic portion of an ecosystem. They are related through the flow of energy and cycling of materials. Each species is identified by its habitat and niche.

Energy passes through the various links, or trophic levels, of a food chain and does not cycle. At each successive link of the chain, only about 10% of the available energy is assimilated into the tissues of the organisms of the next trophic level. Organisms acquire energy through predation and several types of symbiotic relationships: parasitism, commensalism, and mutualism. Different food chains form food webs. The trophic structure of an ecosystem can be summarized by an energy pyramid.

In contrast to energy, inorganic substances such as carbon cycle through the biotic and abiotic parts of an ecosystem. The abiotic component serves as a reserve source and exchange pool for the cycling of elements. Human activity has greatly influenced the effective cycling of many elements.

VOCABULARY

The list in the first column includes some of the chapter's terms. Each term is followed by the page where it first appears. Locate each term in the chapter and read its description. Then match the meanings to the terms.

1. biome (p. 332)
2. carnivore (p. 333)
3. decomposer (p. 333)
4. habitat (p. 333)
5. herbivore (p. 333)
6. mutualism (p. 336)
7. niche (p. 333)
8. producer (p. 333)
9. succession (p. 332)
10. trophic (p. 337)

a. feeding level
b. sequence of communities
c. where organisms live
d. profession of organism
e. organism of decay
f. symbiotic relationship
g. autotrophic organism
h. animal feeding on plants
i. animal feeding on animals
j. terrestrial community

Answers: 1. j 2. i 3. e 4. c 5. h 6. f 7. d 8. g 9. b 10. a

LEARNING ACTIVITIES

Study the text section by section as you answer the following questions.

The Biosphere (p. 332)

1. Select the correct statements.
   a. Temperature and rainfall determine the characteristics of a biome.
   b. Solar energy drives the energy cycle.
   c. Rain is heaviest at the poles and tapers off at the equator.
   d. Climax communities are found in the biosphere.
   e. Tropical rain forests occur where rainfall is scarce.

An Ecosystem (p. 333)

2. Use the letters a–d to indicate the order in which the sun's energy passes through the biotic components of a food chain.

   _____ carnivores
   _____ decomposers
   _____ herbivores
   _____ producers

3. Outline a food chain that is appropriate to the land.
4. In the food chain that you constructed in #3, list the organism(s) that is (are) the:
   a. producer. _____
   b. consumer. _____
   c. herbivore. _____
   d. carnivore. _____
   e. decomposer. _____
   f. autotroph. _____
   g. heterotroph. _____
   h. predator. _____
5. a. Draw a pyramid of energy that indicates the appropriate trophic levels for each member of the food chain.

b. Why does the size of each trophic level decrease from the bottom to the top of the pyramid?

6. Name the type of symbiosis described in each below.
   a. A virus inhabits the human body, causing a disease. _____
   b. A bacterium lives on the skin of a mammal, gaining a habitat but not causing harm to the mammal. _____
   c. Termites live in the intestinal tract of a protozoan, digesting cellulose. _____

Chemical Cycling (p. 339)

7. Select the correct statements about the carbon cycle.
   a. Photosynthesis replaces carbon dioxide in the atmosphere.
   b. Living and dead organisms contain organic carbon.
   c. Most fossil fuels were formed during the Carboniferous period.
   d. Destroying forest removes a reservoir to take up carbon dioxide.
   e. Respiration removes carbon dioxide from the atmosphere.
   f. The oceans are a major reservoir of carbon.
8. Select the correct statements about the nitrogen cycle.
   a. By nitrogen fixation, nitrogen gas is reduced to organic compounds.
   b. Nitrification is the production of nitrates.
   c. Ammonia is converted to nitrates in the soil by nitrifying bacteria.
   d. Denitrification is the conversion of nitrate to nitrogen gas.

Answers:
1. a, b, d
2. 3—carnivores, 4—decomposers, 2—herbivores, 1—producers
3. and 4. see text figs. 21.3, 21.4–21.6, p. 338
5. a. see text fig. 21.7, p. 339 b. each has less available energy
6. a. parasitism b. commensalism c. mutualism
7. b, c, d, f
8. a, b, c, d

CRITICAL THINKING QUESTIONS

1. How does the depletion of an element in the abiotic portion of the environment affect its balance?
2. How do you think that human intervention in forested biomes alters the effect of reaching a climax?

ANSWERS:
1. This depletes the reservoir and exchange pool of the cycle, leaving more in the abiotic portion of the cycle and interfering with a balanced cycle.
2. The interventions disturb the biomes, preventing the realization of a climax.

CHAPTER TEST

Indicate whether each of the following statements is
a. true.
b. false.

     _____ 1. Both herbivores and carnivores are producers in a food chain.
     _____ 2. Energy flows through a food chain because it is constantly lost from organic food as heat.
     _____ 3. A biome is a major aquatic community.
     _____ 4. A pyramid of energy is broadest at the bottom and narrowest at the top.
     _____ 5. Generally the third link in a food chain has about one-half the available energy as the second link.
     _____ 6. The habitat is the so-called profession of an organism in a community.
     _____ 7. Photosynthesis incorporates carbon from the atmosphere into biomass.
     _____ 8. Respiration returns carbon to the atmosphere.
     _____ 9. Nitrogen fixation is the return of the element to the atmosphere.
     _____ 10. Denitrifying bacteria convert atmospheric nitrogen into the bodies of organisms.

Answers: 1. b 2. a 3. b 4. a 5. b 6. b 7. a 8. a 9. b 10. b

CHAPTER 22

Population and Environmental Concerns

CHAPTER CONCEPTS

1. Population growth is curtailed by natural forces.
2. An increasing human population has environmental effects.
3. Pollutants cause the degradation of land, air, and water.

CHAPTER REVIEW

Populations—organisms of the same species living in a given area—have a natural rate of increase (r) in size where $I = rN$. This rate is realized when there is an absence of environmental resistance through density-dependent and density-independent factors. In this case a population realizes its biotic potential. However, usually a compromise occurs between the environmental resistance and biotic potential, leading to a carrying capacity for the population. A population can be characterized by its age structure.

The human population is currently in the exponential part of its growth curve. Developing countries are experiencing a demographic transition. Undesirable consequences of rapid population growth include the degradation of land, loss of biological diversity, water pollution, and air pollution. To preserve the quality of human life, resource consumption is a major concern. The control of population size in developing countries and the consumption per individual in developed countries will contribute to that quality.

VOCABULARY

The list in the first column includes some of the chapter's terms. Each term is followed by the page where it first appears. Locate each term in the chapter and read its description. Then match the meanings to the terms.

1. biological diversity (p. 348)
2. biotic potential (p. 346)
3. carrying capacity (p. 346)
4. environmental resistance (p. 346)
5. greenhouse effect (p. 351)
6. ozone (p. 352)
7. photochemical smog (p. 351)
8. slash and burn (p. 348)

a. type of agriculture
b. mixture of 2 pollutants
c. from buildup of carbon dioxide
d. its shield is deteriorating
e. factors opposing growth of population
f. maximum rate of increase possible
g. actual population size
h. being lost in the tropics

Answers: 1. h 2. f 3. g 4. e 5. c 6. d 7. b 8. a

LEARNING ACTIVITIES

Study the text section by section as you answer the following questions.

Population Growth (p. 346)

1. Identify each of the variables in the following equation:

$$r = \frac{b-d}{N}$$

   $r = $ _____
   $b = $ _____
   $d = $ _____
   $N = $ _____

2. Complete the blanks in the following paragraph.
   The growth curve of most populations is (a) _____ shaped. Under ideal conditions, a population demonstrates its (b) _____ , its maximum rate of increase. Factors of the (c) _____ oppose this potential. Due to these factors, a population usually reaches its (d) _____ , the maximum size that the population can support.
3. Name the 3 age groups of a population.
4. Complete the following.
   a. If the birthrate falls below the death rate, the age structure of a population is _____ shaped.
   b. If the birthrate equals the death rate, the age structure of a population is _____ shaped.
5. Select the correct statements.
   a. The growth of the human population is now exponential.
   b. The human population has undergone 5 phases of population growth.
   c. Russia is an MDC.
   d. The growth rate for the MDCs has now stabilized at 0.6%.
   e. The U.S. has a lower growth rate than the world average.
   f. Africa is an LDC.

Consequences of Population Growth (p. 348)

6. Select the correct statements.
   a. Slash-and-burn agriculture has promoted the growth of forests in the tropics.
   b. Tropical rain forests have held more biodiversity than temperate forests.
   c. Pollutants from solid wastes become part of surface and underground water.
   d. Biological magnification is more likely to occur in aquatic food chains.
   e. Cultural eutrophication can produce massive fish kills.
   f. Plastic bottles are biodegradable.
   g. Photochemical smog contains 2 pollutants.
   h. The greenhouse effect results from the accumulation of carbon monoxide in the atmosphere.
   i. CFCs are the reason for the breakdown of the ozone shield.

Answers:
1. $r = $ rate of natural increase, $b = $ birthrate, $d = $ death rate, $N = $ number of individuals
2. a. J b. biotic potential c. environmental resistance d. carrying capacity
3. dependency, reproductive, postreproductive
4. a. urn b. bell
5. a, c, d, f
6. b, c, d, e, g, i

CRITICAL THINKING QUESTIONS

1. Describe the conditions under which the growth of a population can be infinite.
2. How is the exponential growth of a population similar to the effect of compound interest on money saved in a bank?

Answers:
1. If there were an absence of factors offering environmental resistance, a population could achieve its maximum rate of increase.
2. Interest in the bank is paid on both the principal and the interest that is generated from that base amount and added to the base amount. During exponential growth, new individuals are generated from the original base population and the individuals produced from that original group.

CHAPTER TEST

Indicate your answers by circling the letter. Do not refer to the text when taking this test.

1. In the equation $r = b - d/N$, r represents the
   a. birthrate.
   b. death rate.
   c. rate of natural increase.
   d. size of the original population.
2. The _____ of a population is established by the effects of the _____ .
   a. carrying capacity, environmental resistance
   b. environmental resistance, carrying capacity
3. The biotic potential of a population is the maximum rate of increase.
   a. true
   b. false
4. An age structure diagram is bell shaped when the birthrate
   a. equals the death rate.
   b. exceeds the death rate.
   c. is less than the death rate.
5. The human population has undergone _____ phases of population growth.
   a. 2
   b. 3
   c. 4
   d. 5
6. Asia and Africa are classified as
   a. MDCs.
   b. LDCs.
7. A major effect of the destruction of the tropical rain forest is the _____ of biological diversity.
   a. gain
   b. loss
8. Humans are usually the initial consumers in the case of biological magnification of food chains.
   a. true
   b. false
9. Carbon monoxide is one of the reactants of photochemical smog.
   a. true
   b. false
10. Depletion of the ozone layer will block the arrival of ultraviolet rays on the surface of the earth.
    a. true
    b. false

Answers: 1. c 2. a 3. a 4. a 5. b 6. b 7. b 8. b 9. b 10. b